Mass Transfer

Diffusion & Convection

by
D. James Benton

Preface

Mass transfer is similar to heat transfer in many ways. Some of the same differential equations and solutions will work for either one. There are differences, including the fact that there is often no exchange of matter accompanying an exchange of energy in the form of heat. The topics we consider in this text include: diffusion, dispersion, and sorption. We cover both analytical and numerical solutions.

All of the examples contained in this book,
(as well as a lot of free programs) are available at...

https://www.dudleybenton.altervista.org/software/index.html

A word about units... Most of the examples in this text are given in SI units, as these are common in the field of chemical engineering. The practicing scientist/engineer should be comfortable using a variety of units. Arguments over which are preferable and why only waste time and increase entropy.

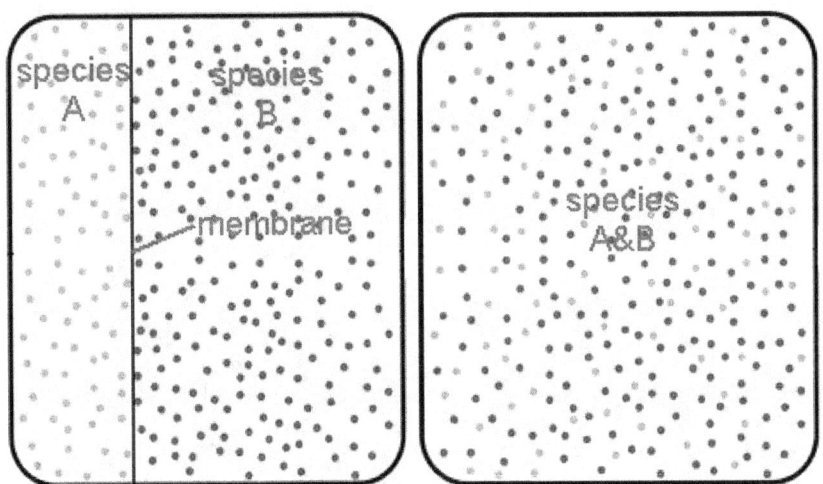

Figure 1. Classic Membrane Problem

Figure 2. Ink Diffusing into Water

Table of Contents

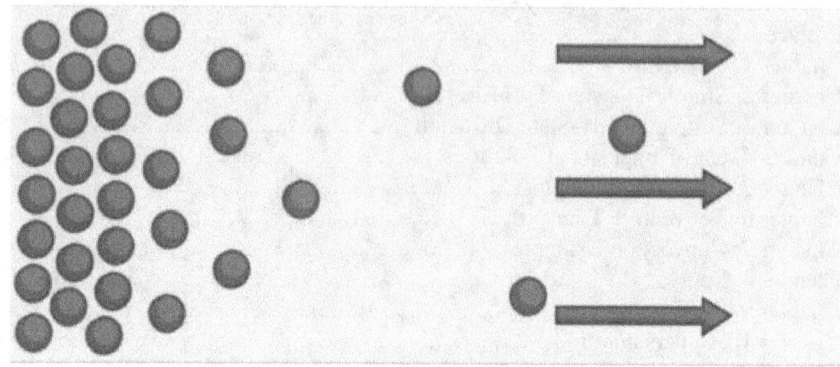

Figure 3. Conceptual One-Dimensional Diffusion

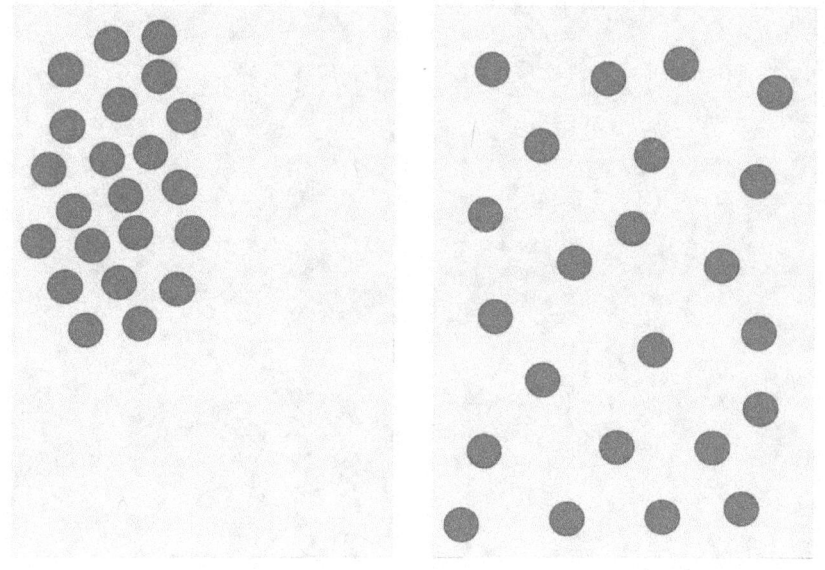

Figure 4. Atoms Expanding to Occupy Available Positions

Chapter 1. Fick's Laws

Mass transfer, as we shall consider it here, is *the movement of matter by virtue of a concentration gradient*. Stating it this way is reminiscent of heat transfer being the *movement of energy by virtue of a temperature gradient*. Mass and energy move in countless flowing situations, but this is not what we are interested in. In this text we will consider those processes *driven* by differences in molecular composition over space and time. The simplest expression of Fick's[1] 1st Law may be stated:

$$J = -D\frac{dC}{dx} \tag{1.1}$$

where J is the flux (mass or moles per unit area per unit time), D is the diffusion coefficient (area per unit time), C is the concentration (i.e., mass or moles per unit volume), and x is the dimension over which the concentration varies (unit length). This is reminiscent of Fourier's[2] 1st law of heat conduction:

$$q = -k\frac{dT}{dx} \tag{1.2}$$

where q is the heat flux (energy per unit area per unit time), k is the thermal conductivity (energy per unit length per unit time per unit temperature), and T is the temperature. By comparing these two differential equations, we should expect that some solutions may apply equally well to both processes.

Fick's 2nd Law arises from the differential of the 1st Law and may be stated for one dimension:

$$\frac{\partial C}{\partial t} = D\frac{\partial^2 C}{\partial x^2} \tag{1.3}$$

which is reminiscent to Fourier's 2nd Law of conduction:

$$\frac{\partial T}{\partial t} = \alpha\frac{\partial^2 T}{\partial x^2} \tag{1.4}$$

where $\alpha = k/\rho C$ is the thermal diffusivity, ρ is the density, and C is the specific heat. We know from heat transfer that Equation 1.4 can be extended to three dimensions and also account for variable properties by:

---

[1] Adolf Eugen Fick (1829–1901) German physicist and physician.
[2] Jean-Baptiste Joseph Fourier (1768–1830) French mathematician and physicist.

$$\frac{\partial T}{\partial t} = \nabla \cdot (\alpha \nabla T) \tag{1.5}$$

where ∇ is the *del operator*. In one-dimensional coordinates (i.e., x) Equation 1.5 becomes:

$$\frac{\partial T}{\partial t} = \frac{\partial}{\partial x}\left(\alpha \frac{\partial T}{\partial x}\right) = \left(\frac{\partial \alpha}{\partial x}\right)\left(\frac{\partial T}{\partial x}\right) + \alpha \frac{\partial^2 T}{\partial x^2} \tag{1.6}$$

We can write similar expressions for mass transfer:

$$\frac{\partial C}{\partial t} = \nabla \cdot (D \nabla C) \tag{1.7}$$

Expanding, we get:

$$\frac{\partial C}{\partial t} = \frac{\partial}{\partial x}\left(D \frac{\partial C}{\partial x}\right) = \left(\frac{\partial D}{\partial x}\right)\left(\frac{\partial C}{\partial x}\right) + D \frac{\partial^2 C}{\partial x^2} \tag{1.8}$$

In three Cartesian coordinates, the del operator is written:

$$\nabla = \frac{\partial}{\partial x} + \frac{\partial}{\partial y} + \frac{\partial}{\partial z} \tag{1.9}$$

In cylindrical coordinates, the del operator is written:

$$\nabla = \frac{\partial}{\partial r} + \frac{1}{r}\frac{\partial}{\partial \theta} + \frac{\partial}{\partial z} \tag{1.10}$$

In spherical coordinates, the del operator is written:

$$\nabla = \frac{\partial}{\partial r} + \frac{1}{r}\frac{\partial}{\partial \theta} + \frac{1}{r \sin \theta}\frac{\partial}{\partial \varphi} \tag{1.11}$$

From these forms of the del operator, we can derive the governing partial differential equation appropriate for a wide variety of problems. To illustrate this we will consider a simplified problem. Suppose a jar containing a substance is opened and allowed to interact with air in the room. We suppose the substance is not volatile so that it does not vaporize upon contact with the air. It simply *diffuses* into the room. We further suppose that there is none of the substance in the air before the jar is opened. This situation would produce a concentration gradient from the jar into the room because the substance exists in the jar at a concentration we assume to be unity (100%) and in the room we assume to be nil (0%). Provided the diffusion coefficient for the substance into air is greater than zero (it can never be negative) and there is some finite, non-trivial distance characterizing the interaction between the substance and the air, we could estimate the flux using Equation 1.1.

Consider the following figure depicting this scenario. Let's assume the jar is 8 cm in diameter. This would make the red plume of diffusing substance rising out of the jar about 12 cm tall. The diffusivity of Para-dichlorobenzene (mothballs) in air is about 0.07 cm²/sec and so we will use this value for our example. Let us also assume a density, ρ, of 1.25 gm/cm³.

Figure 5. Hypothetical Substance Diffusing into Air

Equation 1.1 could be approximated by a finite difference:

$$J = -D\frac{\Delta C}{\Delta x} \qquad (1.12)$$

This yields a flux of (0.07 cm²/sec)*(1-0)*(1.25 gm/cm³)/(12 cm)=0.0073 gm/cm²/s. The area of the jar open to the room is $\pi d^2/4$, making the diffusive mass rate 0.37 gm/sec. If this process continued at the current estimated rate,

3

given the total size of the jar and substance contained ($V=\pi h d^2/4$, $m=\rho V$), all of the substance would diffuse out of the jar and into the air in 1371 sec or about 23 minutes.

Chapter 2. Simple Transient Diffusion

As we mentioned in Chapter 1, the partial differential equation governing transient diffusion is the same as that governing heat conduction. If the process is only one-dimensional and the diffusion coefficient is constant, we can employ the same solution first proposed by Fourier in 1822:

$$C(x,t) = \frac{C_0 \delta}{\sqrt{4\pi Dt}} \exp\left(\frac{-x^2}{4Dt}\right) \tag{2.1}$$

Equation 2.1 gives the concentration, C, as a function of time, t, and distance, x, with δ being some characteristic length. We note several things about this solution. First, at $t=0$, C is unbounded. At $x=0$ the concentration is ∞ at $t=0$ and falls off with $\sqrt{t}$. If we integrate Equation 2.1 from $x=-\infty$ to $x=+\infty$, we get $C_0\delta$, regardless of t, so that mass is conserved. We can plot this equation using an Excel® spreadsheet (example1.xls), which you will find in the examples folder of the online archive accompanying this text. The result is shown in the following figure:

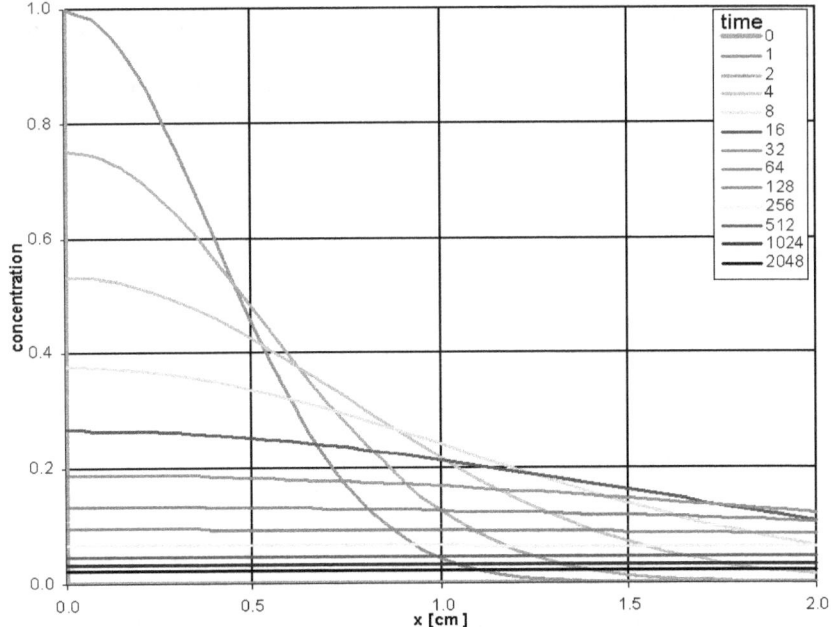

Figure 6. One-Dimensional Transient Diffusion

The magenta line corresponding to *t=0* follows the Y-axis down to zero and then along the X-axis, indicating that the initial concentration is zero everywhere but *x=0*. The top red curve represents the first non-zero time (*t=1*). This falls rapidly as *x* increases, eventually approaching zero in the figure.

Continuum Assumption

This is, of course, *idealized* diffusion, which assumes a *continuum*. Real diffusion involves individual atoms (or molecules). The change from *t=0* to *t=1* appears to be abrupt in the preceding figure, yet we know from experience this couldn't happen so fast. Net migration of the substance into the air cannot occur any more rapidly than the atoms are moving. This molecular scale activity is smeared over as soon as we write a differential equation, because we implicitly presume a continuum by doing so. Our results cannot possibly be any more accurate than the assumption on which the solution rests.

Consider the figures on pages i, ii, iv, and 3. Figures 1, 3, and 4 are conceptual and show individual atoms. Figures 2 and 5 are photographs (albeit modified). These do not exhibit what might be considered a continuous process or one occurring within a continuum. This is particularly evident in Figures 2 and 5, which show multiple length scales (small areas of rapid change and larger areas of nearly uniform concentration).

Apart from molecular-scale modeling, which we will consider later in the text, we must employ mathematical techniques in order to obtain any solutions. This applies equally to analytical or numerical (e.g., finite difference or finite element) methods. For now, we will proceed acknowledging the fact that our assumption of a continuum is not precise and so our solution will not be either.

Chapter 3. Simple Steady-State Diffusion

In this chapter we will eliminate time as a variable by limiting our discussion to steady-state problems. Examples of such processes are few but some may be characterized in this way over a brief period of time. Quiescent, dry air humidifying over a body of water with no significant temperature differential is one example. Although not at all steady-state on a small scale and certainly not adiabatic (no heat transfer), an interesting example of mass transfer is the drinking bird toy depicted below:

Figure 7. Drinking Bird Toy

Dip the bird's beak into the water glass. The red liquid runs along the bird's body toward the head, changing the point of balance. Water on the bill evaporates, providing a slight cooling effect. This causes the liquid to run toward the bird's bottom, changing the point of balance again. The red liquid inside the bird is actually a refrigerant, not merely water. Heat transfer from the air into the bird (as the bird is now slightly cooler than the room), increases the vapor pressure of the refrigerant in the bird's bottom. This forces the liquid up the tube, shifting weight toward the head.

Once started, the bobbing motion will continue until all the water in the glass is gone or the room is saturated with water vapor, which at 15°C and one atmosphere (101.325 kPa), is 1.06% by weight. This is *not* a perpetual motion machine. Entropy is most assuredly being generated. You have provided the relatively dry room and you continue to fill the glass with water. The total potential of the system (bird+water+room) has less potential for doing work after than before, in keeping with the 2nd Law of Thermodynamics. This would be true even if you were to connect a little generator to the bird and even if it continued to drink. A flock of such toys will not solve the energy crisis.

The point of this illustration is that a diffusion process (water moving from the glass into the air via the drinking bird) can continue for quite some time with molecules of air and water moving throughout the room while the concentration in two spatial locations varies only slightly. It will always be 100% at the surface of the water in the glass and initially on the bird's beak after it is dipped. It will always be no more than 1.06% in the room.

There are also many processes where some substance (for example, carbon dioxide or other emissions from a power plant) enters the environment and spreads out into the atmosphere by a combination of diffusion, dispersion, and advection. While these simple examples may seem contrived and trivial, they can provide insight into real world problems.

If we eliminate time as a variable and assume a constant diffusion coefficient, Equation 1.7 becomes:

$$\nabla^2 C = 0 \tag{3.1}$$

This is Laplace's[3] Equation. In Cartesian coordinates, this becomes:

$$\frac{\partial^2 C}{\partial x^2} + \frac{\partial^2 C}{\partial y^2} + \frac{\partial^2 C}{\partial z^2} = 0 \tag{3.2}$$

If we were to somehow maintain one wall at a constant concentration of 100% (e.g., a slab of Para-dichlorobenzene) opposite another wall at a constant concentration of 0% with no significant variation in the other two dimensions (y and z), this problem could be stated:

---

[3] Pierre-Simon, marquis de Laplace (1749–1827) French mathematician, making significant contributions to engineering, mathematics, statistics, physics, astronomy, and philosophy.

$$\frac{\partial^2 C}{\partial x^2} = 0$$
$$C\big|_{x=0} = 1$$
$$C\big|_{x=L} = 0$$

(3.3)

The solution to this problem is indeed trivial:

$$C = \frac{L - x}{L}$$

(3.4)

which is a straight line from the left to the right boundary condition. Solutions become more interesting (and meaningful) when we consider two dimensions:

$$\frac{\partial^2 C}{\partial x^2} + \frac{\partial^2 C}{\partial y^2} = 0$$

(3.5)

One solution to Equation 3.5 is:

$$C = a\left(x^2 - y^2\right)$$

(3.5)

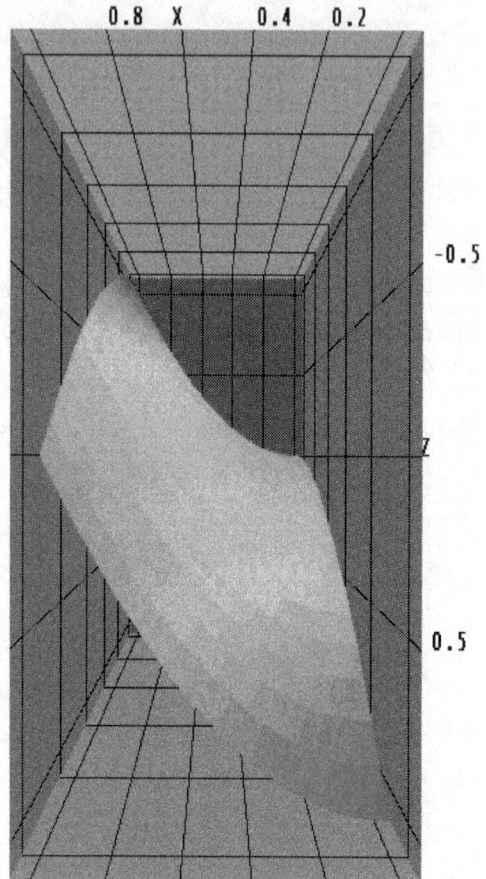

Figure 8. Equation 3.5

While an interesting mathematical solution, this surface does not correspond to any meaningful diffusion problem. A slightly more interesting solution is often presented in the context of solutions to the Laplace equation in two dimensions with constant boundary conditions at x=a and y=b. This solution is found from the general solution and can be stated in the form of an infinite series:

$$C(x, y) = \sum_{n=1}^{\infty} A_n \sin\left(\frac{n\pi x}{a}\right) \sinh\left(\frac{n\pi y}{a}\right)$$

$$A_n = \frac{(-1)^n \left(\dfrac{2}{n\pi}\right)}{\sinh\left(\dfrac{n\pi b}{a}\right)}$$

(3.6)

The calculations can be found in example3.c and the results in example3.tb2 in the examples folder. The surface is shown in this next figure:

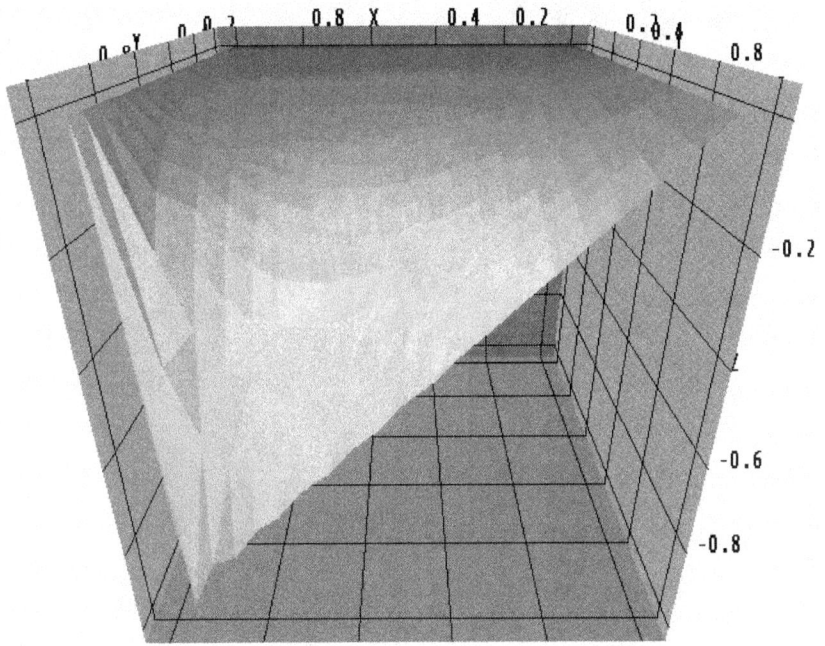

Figure 9. Equation 3.6

Chapter 4. One-Dimensional Solutions

Perhaps the simplest analytical solution that yields near meaningful results that can be used to estimate real world problems is the case of two semi-infinite domains, which share a common interface at $x=0$, one extending to the left ($x<0$) infinitely and the other extending to the right ($x>0$) infinitely. In this case, the solution to Equation 1.3 is:

$$C(x,t) = \frac{1 + erf\left(\dfrac{x}{\sqrt{4Dt}}\right)}{2} \qquad (4.1)$$

where erf() is the Gauss[4] error function, defined by:

$$erf(x) = \frac{2}{\pi} \int_0^x e^{-t^2}\, dt \qquad (4.2)$$

Limiting values include $erf(0)=0$, making $C(0,t)=0.5$; $erf(+\infty)=1$, making $C(+\infty,t)=1$; and $erf(-\infty)=-1$, making $C(-\infty,t)=0$.

The Interface Problem

At $t=0$, all of component A is to the left of $x=0$ and all of component B is to the right of $x=0$. This is illustrated in the following figure:

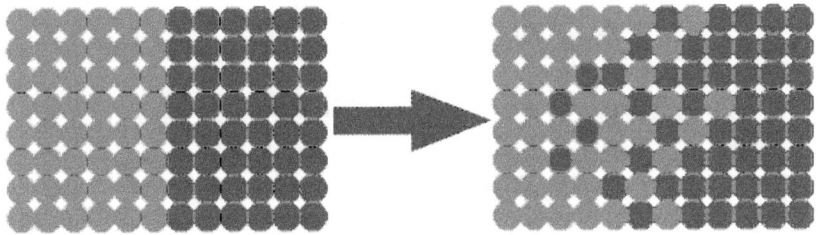

Figure 10. Semi-Infinite Problem at Atomic Level

This problem is illustrated in the spreadsheet example4.xls. While Excel does provide both the error function, erf(), and its complement, erfc(), the form is quite annoying. Excel wants: ERF(lower_limit,upper_limit). No one uses the error function in this way. Should you need integration limits other than 0 and x,

---

[4] Johann Carl Friedrich Gauss (1777–1855) German mathematician and physicist.

you can easily subtract erf(b)-erf(a) to obtain the desired result. The error function is often used with negative arguments, as in this current example. The Excel function as is requires you to swap the arguments, that is, ERF(-x,0) for negative values and ERF(0,x) for positive ones, requiring an additional IF(...) statement and calculation of the argument twice. Rather than implement this wasteful and unnecessary extra step, a macro is used. For details, see Appendix B. The result is shown in the following figure.

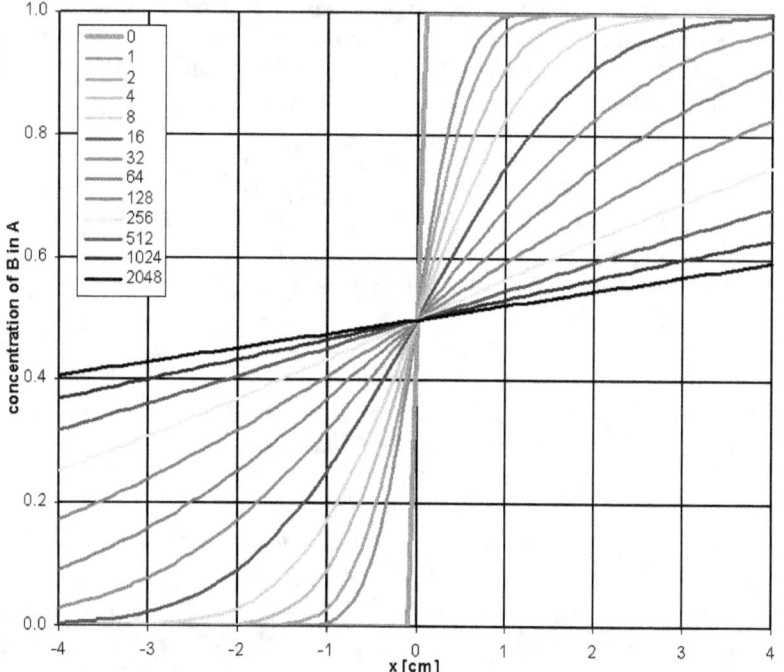

Figure 11. Example 4 – Semi-Infinite Domains

The Y-axis is labeled, "concentration of B in A." This means the substance initially on the right side (*x>0*) of the domain diffusing into the substance initially on the left side (*x<0*) of the domain. This introduces another important discussion: *B* doesn't simply move into and mix with *A*. *A* also moves into and mixes with *B*. Furthermore, the two substances may not be in the same state (i.e., solid, liquid, or vapor). The example we have already mentioned of water diffusing into air started as a liquid and became one vapor within a second vapor. Para-dichlorobenzene at room temperature and pressure is solid. It readily becomes a vapor and diffuses into air with a distinctive odor (mothballs that apparently some insects detest more so than the humans who protect their woolen garments with the smell).

It is often assumed that for two species in the same state, the diffusion coefficient of A into B is the same as B into A.

We see from the preceding figure that the concentration of B in A on the left side increases proportionately to the decrease of B on the right side, as the error function is symmetric. Furthermore, we presume the concentration of A over time and space will be a mirror image of the preceding figure.

The Slab Problem

While the previous geometry (two semi-infinite reservoirs) may approximate some realistic physical problems, most actual situations involve a finite amount of the target species. While the jar problem presented in Chapter 1 might be approximated by two semi-infinite domains for some brief time, in most cases there is far more available volume in the receiving domain than the diffusing species. By superimposing two error function solutions (one at *+b/2* and one at *–b/2*), we arrive at what is often called the slab problem.

$$C(x,t) = \left(\frac{a}{2}\right)\left[erf\left(\frac{x+\dfrac{b}{2}}{\sqrt{4Dt}}\right) - erf\left(\frac{x-\dfrac{b}{2}}{\sqrt{4Dt}}\right)\right] \tag{4.3}$$

where *a* is the initial concentration and *b* is the thickness of the slab. The calculations can be found in spreadsheet example5.xls and the results are shown in the figure below:

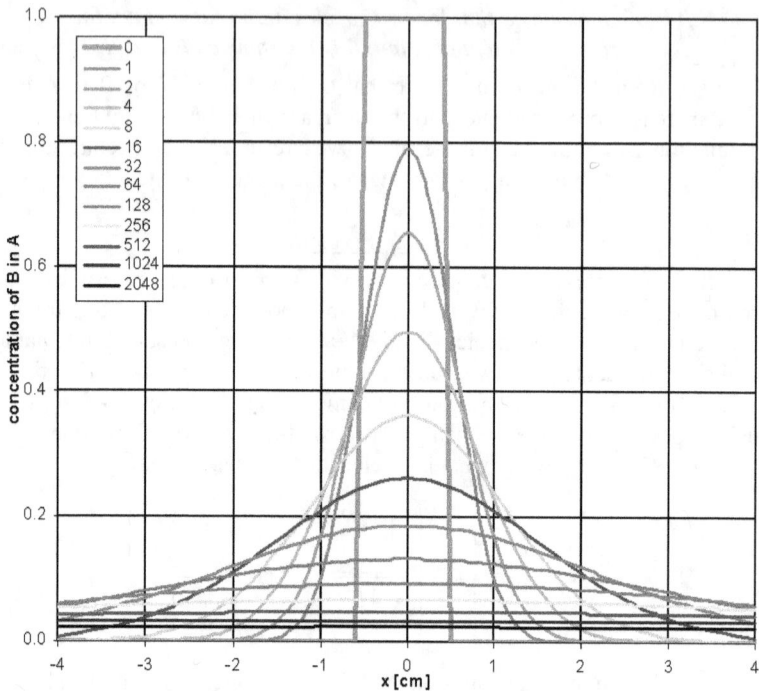

Figure 12. Example 5 - Slab Problem

In the first case (two semi-infinite domains) the concentration at $x=0$ remained constant at 0.5 over time, which would be true of a semi-infinite source. In the second case (the finite slab problem) the concentration at $x=0$ diminishes over time, which is consistent with a finite quantity of the substance. In fact, the decrease over time at $x=0$ is shown in this next figure:

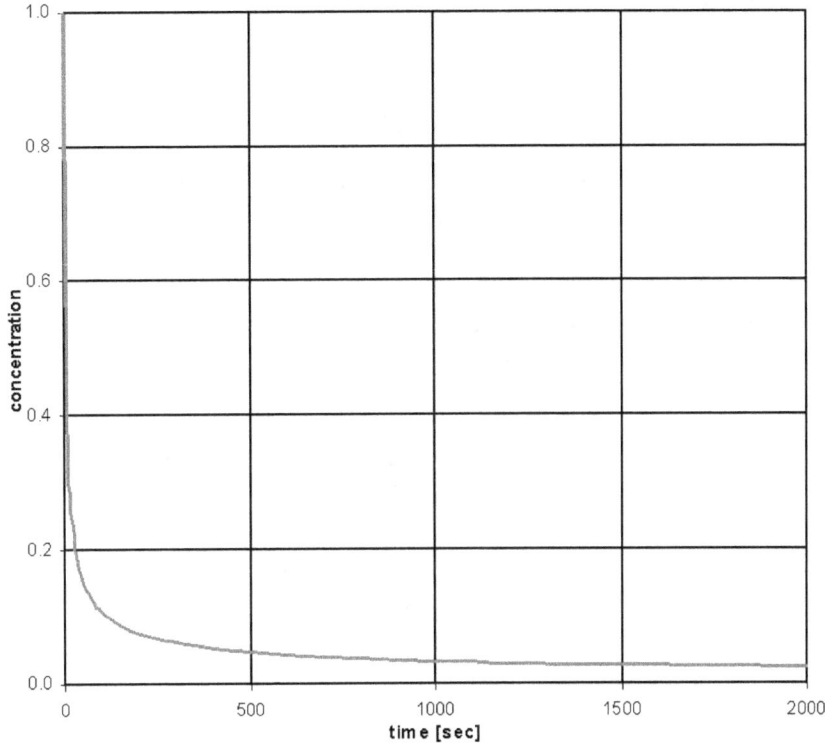

Figure 13. Concentration at x=0

Both of these solutions are transient, one-dimensional, and constant properties in Cartesian coordinates. Next we will consider spherical and cylindrical coordinates, also in one dimension.

1D Cylindrical Problem

Equation 1.7 in one-dimensional cylindrical coordinates (ignoring variation in θ and z) is written:

$$\frac{\partial C}{\partial t} = \frac{1}{r}\frac{\partial}{\partial r}\left(rD\frac{\partial C}{\partial r}\right)$$ (4.4)

General solutions to this partial differential equation are infinite series involving Bessel functions, which are not particularly useful. Instead, we can find approximate solutions. As for many cylindrical problems, we expect that the concentration will fall off with the square of the radius. The partial derivative with respect to r would be proportional to $-2/r$. The flux would then

17

be $J=-2rD/a^2$, where a is the radius at $t=0$. The area of a cylinder is $A=2\pi rh$ and the volume is $V=\pi r^2 h$, making the conservation of mass (or moles):

$$\frac{d(\pi r^2 h)}{dt} = \left(\frac{-2rD}{a^2}\right)(2\pi rh) \qquad (4.5)$$

The length, h, of the cylinder is presumed constant so that the solution to this ordinary, separable differential equation is:

$$r = ae^{\frac{-2Dt}{a^2}} \qquad (4.6)$$

We solve this problem numerically in spreadsheet example6.xls and plot this along with the analytical solution.

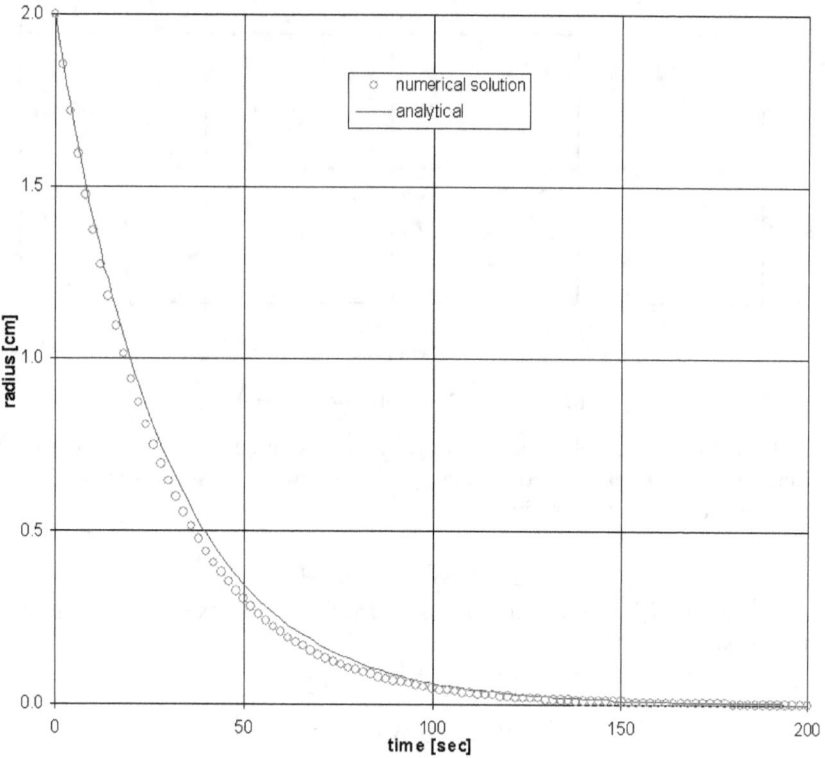

Figure 14. Approximate 1D Cylindrical Solution

1D Spherical Problem

Equation 1.7 in one-dimensional spherical coordinates (ignoring variation in θ and φ) is written:

18

$$\frac{\partial C}{\partial t} = \frac{1}{r^2}\frac{\partial}{\partial r}\left(r^2 D \frac{\partial C}{\partial r}\right) \tag{4.7}$$

General solutions to this partial differential equation are also infinite series, which are also not particularly useful. Instead, we can find approximate solutions. As for many spherical problems, we expect that the concentration will fall off with the cube of the radius. The partial derivative with respect to r would be proportional to $-3/r^2$. The flux would then be $J=-3r^2D/a^3$, where a is the radius at $t=0$. The area of a sphere is $A=r\pi r^2$ and the volume is $V=4/3\pi r^3$, making the conservation of mass (or moles):

$$\frac{d\left(\dfrac{4\pi r^3}{3}\right)}{dt} = \left(\frac{-3r^2 D}{a^3}\right)\left(4\pi r^2\right) \tag{4.8}$$

The solution to this ordinary, separable differential equation is:

$$r = \frac{a}{\left(\dfrac{3Dt}{a^2}\right)+1} \tag{4.9}$$

Note that the first term in the numerator of Equation 4.9 is non-dimensional so that the final result has the dimensions of length. We also solve this problem numerically in spreadsheet example7.xls and plot this along with the analytical solution. The results are shown in the following figure.

19

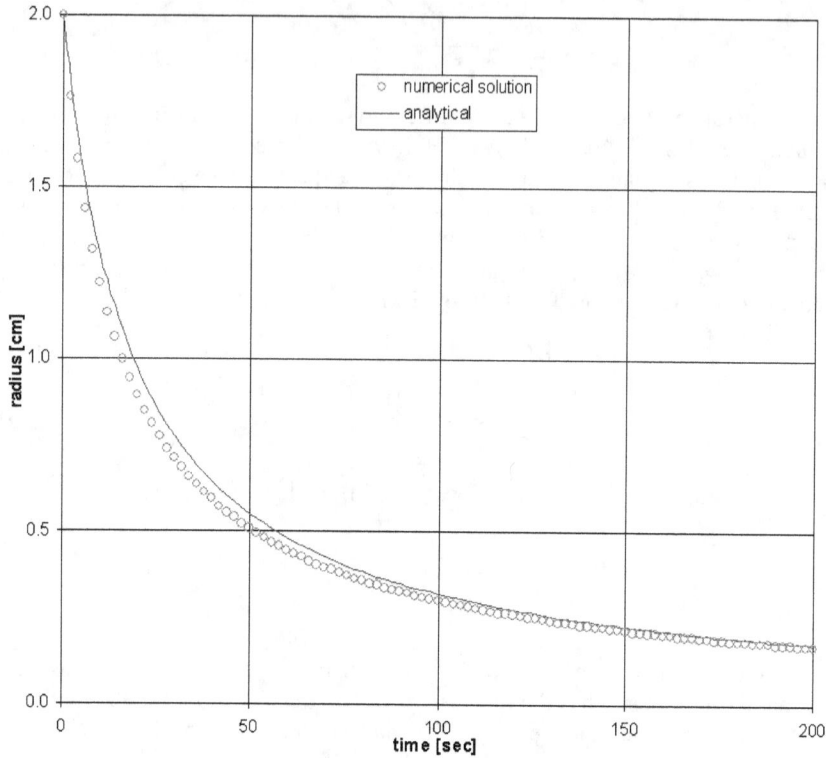

Figure 15. Approximate 1D Spherical Solution

Comparing this figure to the previous one, we see that the expected time for a substance in cylindrical shape to mostly diffuse into the surrounding media is much more rapid than for the same substance in spherical shape. This should not be surprising considering the differing relationship between surface area and volume. Diffusive transfer is proportional to area, while mass is proportional to volume. This difference is worth bearing in mind when considering the deployment. If you wanted to stretch out the process, spheres would be the preferred choice over rods. That's precisely why they're called moth*balls*. The expectation is for them to protect against insects over extended periods.

Chapter 5. Diffusion Examples

According to a study released by the National Institutes of Health[5], "a single box of mothballs containing 396 g of naphthalene, released within an indoor residential environment, is capable of raising the indoor air concentration to an average of approximately 200 µg/m³ for a period of one year." What volume of air would contain this amount at this concentration?

$$V = \left(\frac{396g}{200\frac{\mu g}{m^3}}\right)\left(\frac{10^6 \mu g}{g}\right) = 1,980,000m^3 \qquad (5.1)$$

If the rooms in this "residential environment" were 2 m tall, what would the floor plan area have to be?

$$A = \frac{V}{h} = \frac{1,980,000m^3}{2m} = 990,000m^2 = 0.99km^2 \qquad (5.2)$$

a large "house" indeed! This is two-thirds the volume of Super Bowl Stadium. Clearly, most of the naphthalene leaked out of the residence or chemically degenerated (oxidized) over this time. If the density of naphthalene solid were 1.145 g/cm³, what size sphere would be 396 g?

$$r = \left(\frac{3m}{4\pi\rho}\right)^{\frac{1}{3}} = \left[\frac{3(396g)}{4\pi\left(\frac{1.145g}{cm^3}\right)}\right]^{\frac{1}{3}} = 4.35cm \qquad (5.3)$$

The volume in Equation 5.1 is equivalent to a sphere of what radius?

$$r = \left(\frac{3V}{4\pi}\right)^{\frac{1}{3}} = \left[\frac{3(1,980,000m^3)}{4\pi}\right]^{\frac{1}{3}} = 77.9m \qquad (5.4)$$

---

[5] Sudakin, D. L., Stone, D. L., and Power, L., "Naphthalene Mothballs: Emerging and Recurring Issues and Their Relevance to Environmental Health," Current Topics in Toxicology, Vol. 7, pp 13–19, 2011 https://www.ncbi.nlm.nih.gov/pmc/articles/PMC3850774/

If the naphthalene spread outward at a rate of $r=\sqrt{4Dt}$, what would the diffusion coefficient have to be to extend from near zero (by comparison) to 77.9 m in one year?

$$D = \frac{r^2}{4t} = \frac{\left[(77.9m)\left(\dfrac{100cm}{m}\right)\right]^2}{4(1y)\left(\dfrac{365d}{y}\right)\left(\dfrac{24h}{d}\right)\left(\dfrac{3600s}{h}\right)} = 0.481\frac{cm^2}{s} \qquad (5.5)$$

Figure 16. Mothballs as Sold Today

The measured diffusion coefficient of naphthalene in air is about 0.059 cm²/s or about 12% of this value. It would seem from this calculation that most (perhaps 88%) of the naphthalene would have to chemically degenerate

22

(oxidize), as it couldn't possibly expand via diffusion to such a volume indicated by this small concentration.

Figure 17. Mothballs on Grass to Provide Scale

Using the "slab" solution Equation 4.3, what concentration might one expect at 77.9 m distance after diffusion for one year at 0.059 cm²/s? The calculations and results may be found in example8.xls. The results are shown below:

	A	B
1	user inputs	
2	1	t [years]
3	4.35	b/2 [cm]
4	77.9	x [m]
5	1.145	a [g/cm³]
6	0.059	D [cm²/s]
7	calculations	
8	365	t [days]
9	31,536,000	t [s]
10	7790	x [cm]
11	5.93E-07	C [g/cm³]
12	0.593	C [μg/cm³]
13	592,810	C [μg/m³]

Figure 18. Example 8 Calculations

23

The calculated concentration is 592,810 μg/m³, almost 3000 times the concentration mentioned in the report. This too would seem to indicate that most of the naphthalene chemically degenerates in a year. We can use this same spreadsheet to calculate the expected concentration at 77.9 m over time. If shown in μg/cm³ and spreading distance in m, both can be displayed on the same scale.

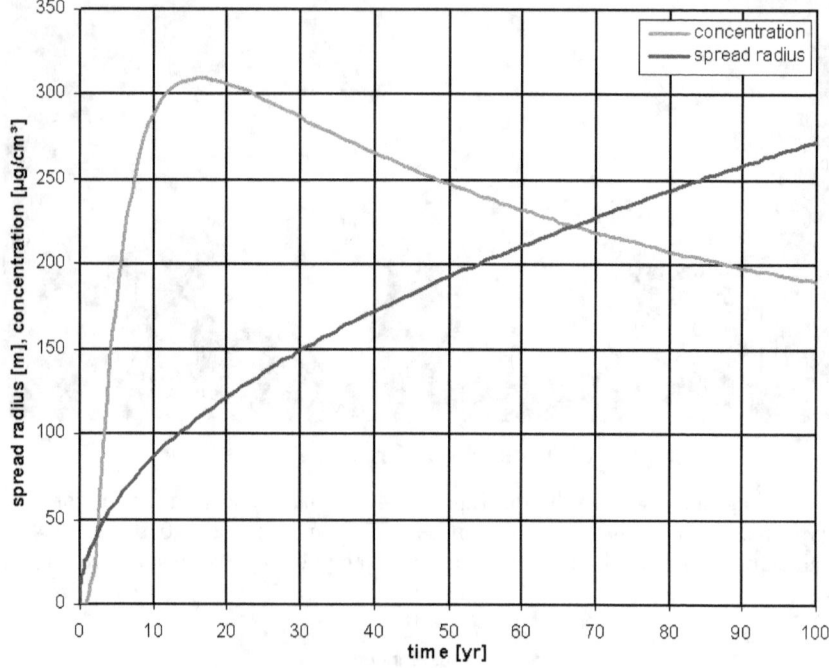

Figure 19. Example 8 Results

We see in the figure that the concentration at this distance (and every other distance by extension) reaches a peak at some time and then diminishes, ultimately reaching 0 at $t=\infty$, while the radius of spread continues to increase with time.

Solvent vs. Solute

In this example we considered a relatively small amount of naphthalene diffusing into a much larger space filled with air. This small/large or dilute/ concentrated disparity is often the case when discussing mass transfer, especially diffusion. We often refer to the receiving fluid as the *solvent* and the diffusing substance as the *solute*.

Diffusion in Liquids

Chlorine is often added to pools via 3-inch tablets. These are said to slowly dissolve over several days. The tablets are sold by the bucket, as shown in the following figure:

Figure 20. Pool Chlorine Tablets

The target level of chlorine is 1 to 2 ppm (parts per million). Clearly, this is far below the saturation limit, as water has a high affinity for free chlorine as well as ionic chloride. For the purposes of illustration, we will assume that migration into the water is by diffusion. If the tablets are 1 inch thick and mostly dissolve (0.5% left) in 3 days (72 hours), what is the diffusion coefficient? We will use Equation 4.3 to solve this problem. The calculations may be found in example9.xls. We seek the value of D such that at $t=72$ hours and $b=2.54$ cm, the average concentration fro $-b/2$ to $+b/2$ is equal to 0.5%.

While we could integrate from $-b/2$ to $+b/2$ and solve this problem with considerable effort, we can also take advantage of a simplification. In the bottom of spreadsheet example5.xls, where we generated Figures 12 and 13, we can also calculate the average and compare this to the central value at $x=0$. That graph (and the following figure) are at the bottom.

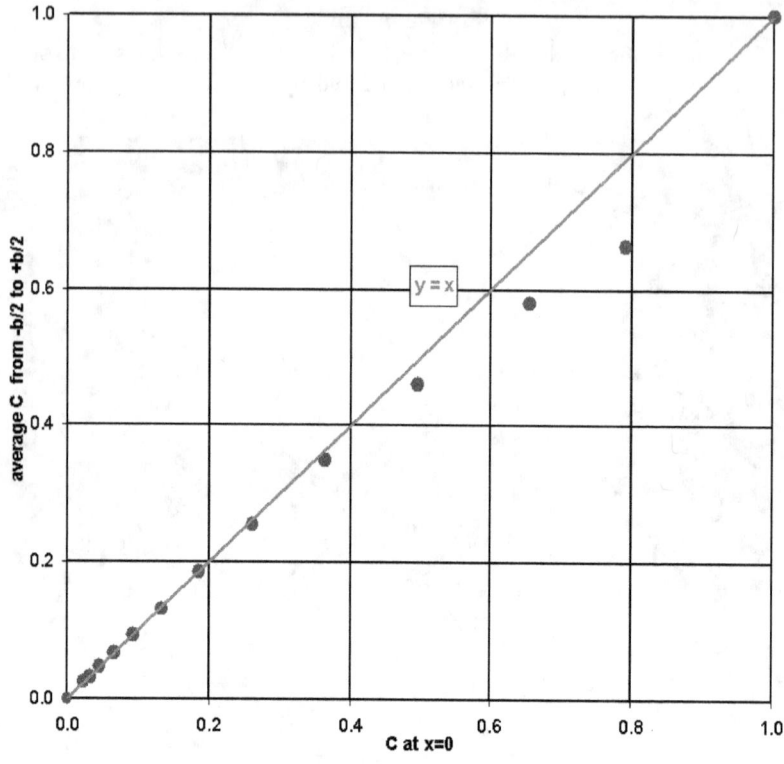

Figure 21. Average vs. Central Value

For results less than 50%, the central and average values are close enough for approximate calculations such as these. The result is 0.079 cm²/s.

	A	B
1	user inputs	
2	**0.079**	D [cm²/sec]
3	2.54	b [cm]
4	calculations	
5	t	mass
6	[hours]	%
7	1	4.24%
8	2	3.00%
9	3	2.45%
76	70	0.51%
77	71	0.50%
78	72	0.50%

Figure 22. Example 9 - Calculations

26

The remaining mass over time is shown in this next figure:

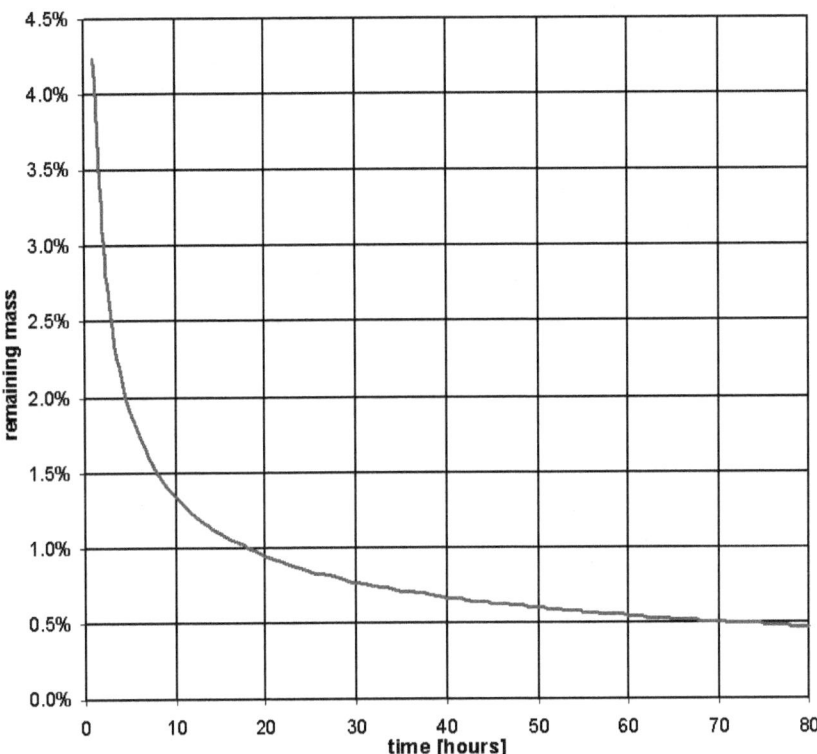

Figure 23. Example 9 - Remaining Mass

Diffusion in Solids

In the manufacture of semiconductors, various substances are diffused into sections of a silicon wafer in order to modify the electrical properties. The process is called *doping*. Arsenic and phosphorus are two common n-type (negative) dopants. Boron and aluminum are common p-type (positive) dopants. The diffusion coefficient and consequent rate of diffusion are temperature-dependent and often elevated temperatures are used to effect infusion of the doping substances.

Using the coefficients in Table A4 and assuming one-dimensional transient diffusion (Equation 2.1) of phosphorus gas into crystalline (solid) silicon , what is the effective penetration dept in 15, 30, 45, and 60 minutes at 2000°C up to 2600°C? The calculations may be found in example10.xls and are shown in the following figure:

27

	A	B	C	D	E	F	G
1	D of P in Si [cm²/s]			effective penetration depth [μm]			
2	temp.	slope	-8.38	duration [minutes]			
3	[°C]	intercept	19.66	15	30	45	60
4	2000	4.40	6.4E-18	0.00152	0.00215	0.00263	0.00304
5	2100	4.21	2.3E-16	0.00907	0.0128	0.0157	0.0181
6	2200	4.04	6.1E-15	0.0469	0.0663	0.0813	0.0938
7	2300	3.89	1.3E-13	0.214	0.302	0.370	0.427
8	2400	3.74	2.1E-12	0.868	1.23	1.50	1.74
9	2500	3.61	2.8E-11	3.19	4.51	5.52	6.37
10	2600	3.48	3.2E-10	10.7	15.1	18.5	21.4

Figure 24. Example 10 - Estimated Penetration Depth

We see a very strong impact of temperature even over this relatively short duration. Temperature control is critical in the manufacture of semiconductors.

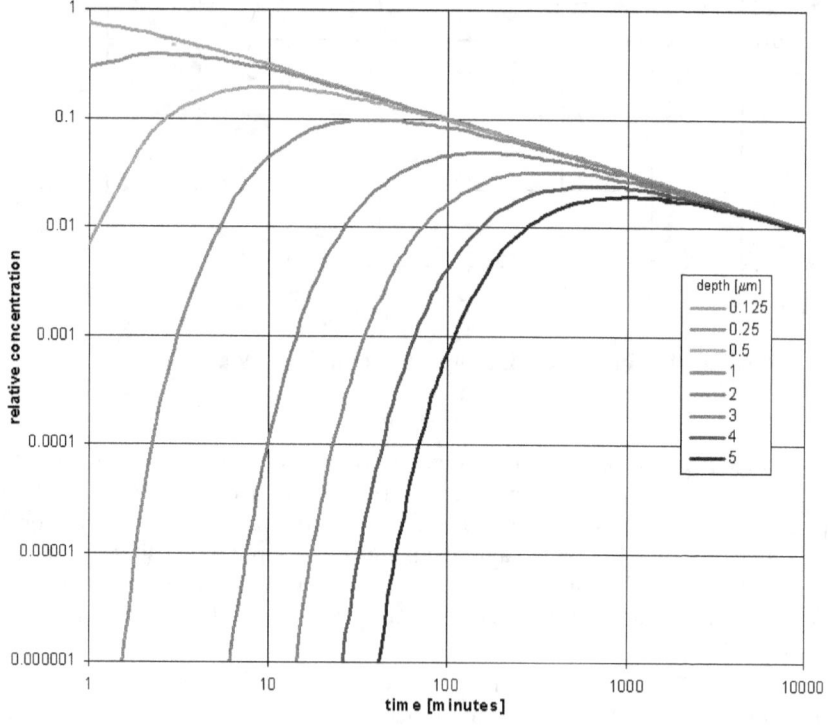

Figure 25. Relative Concentration at Depth for 2400°C

We will consider the temporal (time) and spatial (x) variability of the concentration at 2400°C, just before the large increase in penetration rate. The product of the initial concentration, C_0, and characteristic depth, δ, in

28

Equation 2.1 can be selected so that the limiting value as we approach $t=x=0$ is unity (i.e., 1.0 in the upper left corner of the preceding figure). In the manufacturing process, we are very concerned with concentration, while in this example we will only consider relative values. The product $C_0\delta$ equal to 4×10^{-5} cm or (0.4 μm) will suffice.

In this figure we see that the concentration (relative or absolute) rises over time to some maximum value and then falls off. This behavior arises from two things: 1) increased penetration with time and 2) a constant amount of total mass spread out over a larger volume. If we want to achieve a certain concentration at a certain depth, we must control both time and temperature. There are also bounds on what can be achieved. For example, at this temperature (2400°C), we can never achieve a relative concentration of 0.1 (10%) of phosphorus at a depth of 5 μm in silicon because the dark blue (lowest) curve never reaches this value.

Chapter 6. Convective Transport

Far more experiments have been conducted to study heat transfer than mass transfer. As a result, there are many more empirical correlations and solutions available in the literature for heat than mass transfer. There are far more textbooks and courses available too. Fortunately, there is a relationship between these two so that we can use many of the developments with very little effort and a reasonable degree of confidence.

Reynolds Analogy

We begin with an observation of Osborn Reynolds, who was studying boilers and condensers in the late 19th century. He made considerable use of dimensionless parameters in his quest to estimate the performance of these vital heat exchangers. He noticed a correlation between pressure drop and heat transfer. He found that the heat transfer was proportional to pressure drop, at least for turbulent flow in pipes. After seeing this same proportionality in other devices, arrangements, and conditions, he generalized this relationship into what is called Reynolds[6] Analogy.

$$St = \frac{f}{2} \tag{6.1}$$

where f is the friction factor and St is the Stanton[7] number, which can be related to the Nusselt[8] number, Nu, Reynolds number, Re, and Prandtl[9] number, Pr:

$$St = \frac{Nu}{RePr} = \frac{h}{\rho CV} \tag{6.2}$$

$$Re = \frac{\rho Vd}{\mu} \tag{6.3}$$

---

[6] Osborne Reynolds (1842–1912) pioneer in the field of fluid dynamics, conducting studies of boilers and condensers.

[7] Thomas Edward Stanton (1865-1888) British engineer; studied under Osborne Reynolds, jointly conducted many experiments.

[8] Ernst Kraft Wilhelm Nußelt (1882–1957) German engineer and professor, who performed many studies advancing the science of heat transfer.

[9] Ludwig Prandtl (1875–1953) German engineer and aerodynamicist; pioneer in the field of aeronautics.

$$Pr = \frac{\mu C}{k} \tag{6.4}$$

$$Nu = \frac{hd}{k} \tag{6.5}$$

where C is the specific heat, d is the pipe diameter or characteristic length, k is the thermal conductivity, h is the heat transfer coefficient, V is the velocity, μ is the dynamic viscosity, and ρ is the density.

Friction Factor

The friction factor for flow in a pipe is defined by the Darcy-Weisbach[10,11] equation:

$$\Delta P = f\left(\frac{L}{d}\right)\left(\frac{\rho V^2}{2g}\right) \tag{6.6}$$

where ΔP is the pressure drop, f is the friction factor, L is the length, d is the diameter, ρ is the density, V is the velocity, and g is the gravitational constant. This is most often shown in what is called a Moody[12] chart, after the researcher who collected much experimental data on the subject.

---

[10] Henry Philibert Gaspard Darcy (1803–1858) French engineer who made important advances in the study of hydraulics and flow in porous media.
[11] Julius Ludwig Weisbach (1806-1871) German mathematician and engineer.
[12] Lewis Ferry Moody (1880–1953) American engineer and inventor; the first Professor of Hydraulics in the School of Engineering at Princeton.

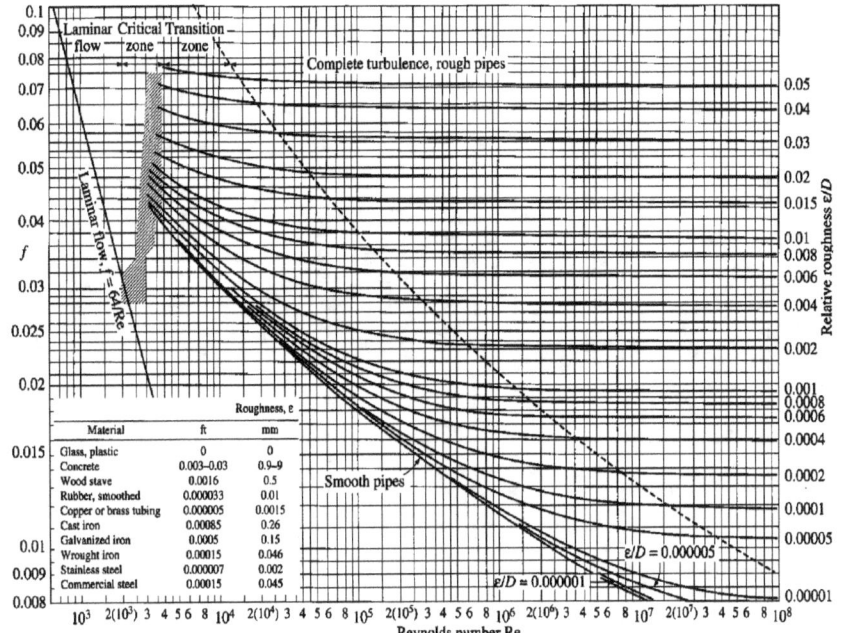

Figure 26. Moody Chart

There are two types of such charts found in the literature. The first is associated with the Darcy-Weisbach formula and the second is associated with the Fanning[13] formula. The only difference is a factor of 4. You can easily tell them apart by closely examining the laminar line on the left side of the chart. If it says *f=Re/64*, then it's Darcy-Weisbach. If it says *f=Re/16*, then it's Fanning. The smaller of the two (Darcy-Weisbach) is used with Reynolds Analogy.

The Colebrook-White[14] empirical relationship for friction factor in pipes of varying roughness, *ε/d*, is given by the following equation and agreement shown in the following figure:

$$\frac{1}{\sqrt{f}} = -2\log_{10}\left(\frac{\varepsilon}{3.7d} + \frac{2.51}{\text{Re}\sqrt{f}}\right) \tag{6.7}$$

---

[13] John Thomas Fanning (1837–1911) American architect and hydraulic engineer who studied pressure loss in pipes.

[14] Colebrook, C. F. and White, C. M., "Experiments with Fluid Friction in Roughened Pipes". Proceedings of the Royal Society of London, Series A, Mathematical and Physical Sciences, Vol. 161, No. 906, pp. 367–381, 1937.

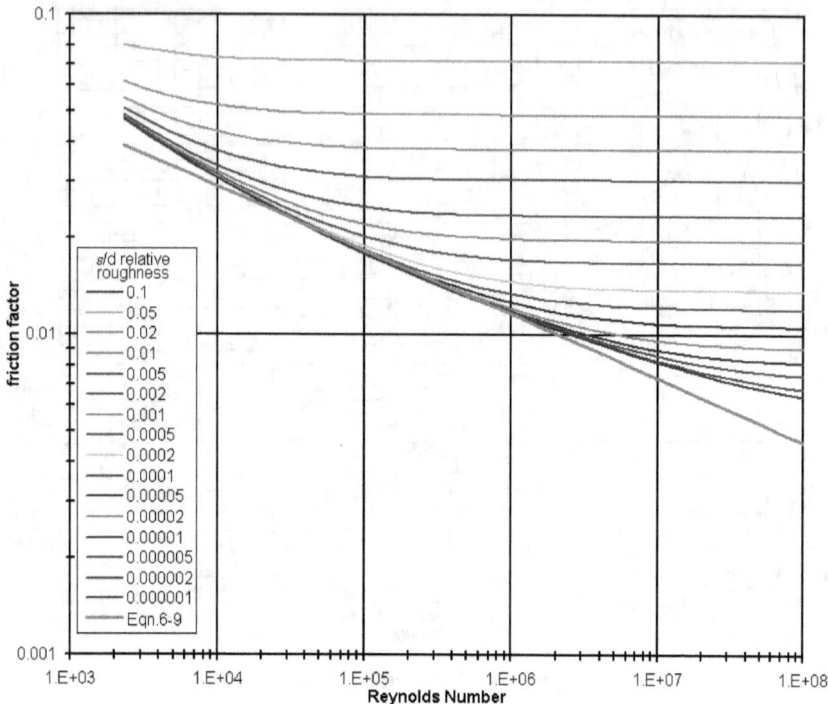

Figure 27. Friction Factor Using Colebrook-White Formula

<u>Heat Transfer for Turbulent Flow in Pipes</u>

Heat transfer for turbulent flow inside pipes can be accurately predicted using the correlation of Dittus-Boelter:[15]

$$Nu = 0.023\,Re^{0.8}Pr^{n} \qquad (6.8)$$

where n=0.4 for heating and n=0.3 for cooling. The range of validity for this empirical correlation is $Re \geq 10^5$ and $0.6 \leq Pr \leq 160$, which is quite broad. Substituting backward, we can infer a friction factor:

$$f = \frac{0.046}{Re^{0.2}Pr^{1-n}} \qquad (6.9)$$

---

[15] Dittus, F. W. and Boelter, L. M. K., "Heat Transfer in Automobile Radiators of the Tubular Type," *Publications in Engineering*, University of California, Berkeley, Vol. 2, 1930.

Ignoring the Prandtl number, this is close to the Blasius[16] correlation for turbulent flow in smooth pipes. Equation 6.9 is shown in the previous figure (thick red line at the bottom) relative to the other curves for rough pipes.

$$f = \frac{0.3164}{Re^{0.25}} \tag{6.10}$$

The Chilton-Colburn Analogy

Inspired by Reynolds Analogy, Chilton[17] and Colburn[18] proposed the following relationship:[19]

$$J_M = \frac{f}{2} = \frac{Sh}{ReSc^{\frac{1}{3}}} \tag{6.11}$$

where J_M is the Chilton-Colburn dimensionless mass transfer factor (equal to $f/2$), Sh is the Sherwood[20] number, and Sc is the Schmidt[21] number. The dimensionless Sherwood number is the ratio of the mass transfer coefficient to the rate of diffusion

$$Sh = \frac{Kd}{D} \tag{6.12}$$

where K is the mass transfer coefficient, d is the characteristic length (e.g., pipe diameter), and D is the diffusion coefficient. Note that sometimes the symbol h is used instead of K. As we discuss both heat and mass transfer in this text, we will use the two different symbols. The Schmidt number is defined by:

$$Sc = \frac{\mu}{\rho D} \tag{6.13}$$

where μ is the dynamic viscosity, ρ is the density, and D is the diffusion coefficient.

---

[16] Paul Richard Heinrich Blasius (1883–1970) a German physicist who studied fluid dynamics under Prandtl.

[17] Thomas H. Chilton (1899–1972) professor and founder of modern chemical engineering.

[18] Allan Philip Colburn (1904-1955) chemical engineer and humanitarian.

[19] Chilton, T. H. and Colburn, A. P., "Mass Transfer (Absorption) Coefficients Prediction from Data on Heat Transfer and Fluid Friction," Industrial & Engineering Chemistry, Vol. 26, No. 11, pp. 1183-1187, 1934.

[20] Thomas Kilgore Sherwood (1903–1976) American chemical engineer and a founding member of the National Academy of Engineering.

[21] Ernst Heinrich Wilhelm Schmidt (1892–1975) German engineer, developer of superheated steam technology and properties.

Chapter 7. Convection Examples

Ethyl acetate (CH_3–COO–CH_2–CH_3) is an organic compound used in some glues and nail polish removers. It is also used in the decaffeination process of tea and coffee [that's disturbing]. In one such process, employing a tubular exchanger, the device must be periodically swept of residue by forced air. The mean velocity of air in the tubes is 4 m/s and the temperature is 25°C. The inside diameter of the tubes is 3 cm. Estimate the mass transfer coefficient and flux based on the Chilton-Colburn Analogy.

From various sources on the Web (e.g., www.engineeringtoolbox.com), we find the properties of air at these conditions. The density, ρ, is 1.25 kg/m³, and the dynamic viscosity, μ, is 0.0172 cP (0.0000172 kg/m/s). The Reynolds number is:

$$Re = \frac{\rho V d}{\mu} = \frac{\left(1.25\frac{kg}{m^3}\right)\left(4\frac{m}{s}\right)(0.03m)}{\left(0.0000172\frac{kg}{ms}\right)} = 8721 \qquad (7.1)$$

We check to be sure that this is in the turbulent range (>4000). As the inside of the tubes are covered with ethyl acetate, we can assume the effective surface is smooth (i.e., $\varepsilon/d=0$). Using the Colebrook-White equation for friction factor (see spreadsheet friction_factor.xls in the examples folder) we get f=0.032. As the flow is predominantly air, we use air properties to evaluate the Reynolds number and friction factor. From Equation 6.11 we get J_M=0.016. The diffusion coefficient of ethyl acetate into air (which we presume is the same as air into ethyl acetate) is 0.085 cm²/s (0.0000085 m²s). The Schmidt number is then:

$$Sc = \frac{\mu}{\rho D} = \frac{\left(0.0000172\frac{kg}{ms}\right)}{\left(1.25\frac{kg}{m^3}\right)\left(0.0000085\frac{m^2}{s}\right)} = 1.62 \qquad (7.2)$$

We then get the Sherwood number from Equation 6.11:

$$Sh = \frac{f}{2} Re Sc^{\frac{1}{3}} = \left(\frac{0.032}{2}\right) \times 8721 \times 1.62^{\frac{1}{3}} = 164 \qquad (7.3)$$

We get the mass transfer coefficient, K, from Equation 6.12:

$$K = \frac{ShD}{d} = \frac{164\left(0.0000085\dfrac{m^2}{s}\right)}{(0.03m)} = 0.0465\frac{m}{s} \qquad (7.4)$$

As the concentration of ethyl acetate coating the inside of the pipe is 100% and the air is initially free of this substance, the flux is numerically equal to this same value, or 0.0465 moles/s/m². The transfer area for 1 m of tube length is $A=\pi dL$ or 0.0942 m², making the molar flux equal to 0.0438 moles/s. The molecular weight of ethyl acetate is 88.11 g/mole, making the mass flux equal to 0.386 g/s/m/tube. For 100 tubes the mass transfer is 2.315 kg/min.

What if the mean velocity of air were reduced to 3 m/s? The Reynolds number would be 6541 (still turbulent) and the friction factor would be 0.0346. The Sherwood number would be 133. The flux would then be 0.0377 moles/s/m² or 0.313 g/s/m/tube or 1.877 kg/min.

What if the mean velocity of the air were increased to 5 m/s? The Reynolds number would be 10,901 (fully turbulent) and the friction factor would be 0.0315. The Sherwood number would be 116. The flux would then be 0.0328 moles/s/m² or 0.273 g/s/m/tube or 1.635 kg/min.

External Flow

We have seen how internal flows (e.g., inside a pipe) might be analyzed. Many mass transfer processes occur on outside surfaces. The simplest conceptual situation would be flow over a flat plate and for that we have correlations readily available. As the Reynolds number changes along the plate, we will consider several points, rather than several velocities, as in the previous example.

Hydrogen sulfide (H_2S, molecular weight 34.1) is an undesirable product of many reactions, including the work of anaerobic bacteria. It is toxic in concentration and smells like rotten eggs. Consider a process that produces unwanted H_2S, which is carried away by a stream of water at 35°C. The saturation limit of H_2S in water is equivalent to about 0.1 molar solution or 3.41 g per kg (or liter) of water. The capacity of air to contain H_2S is much higher than this value so that we can consider it unlimited by comparison.

The water is spreads out over flat surfaces over which air flows at a mean velocity of 1.25 m/s. Hydrogen sulfide is removed from the water into the air via convective transport. We first calculate the Reynolds number and friction factor at three locations: $x=0.5$ m, 5 m, and 50 m. From various sources on the Web (e.g., www.engineeringtoolbox.com), we find the properties of air at these conditions. The density, ρ, is 1.145 kg/m³, and the dynamic viscosity, μ, is 0.01805 cP (0.00001805 kg/m/s).

Replacing d with x in Equation 6.3, we calculate the results in spreadsheet example11.xls. The resulting three Reynolds numbers are $4x10^4$, $4x10^5$, and $4x10^6$. The flow is presumed laminar for $Re<10^5$ and turbulent for $Re>3x10^6$ so that our three points are in the laminar, transition, and turbulent regions. In the laminar region, the local friction factor (at a specific value of x) is given by:

$$f = \frac{0.664}{\sqrt{Re}} \qquad (7.5)$$

In the turbulent region, the local friction factor is given by:

$$f = \frac{0.059}{Re^{0.2}} \qquad (7.6)$$

There is no standard formula for the transition region, as it is somewhat unpredictable. For the purposes of illustration here, we will simply use the square root of the product given by the two equations evaluated at the transitional Reynolds numbers (see example11.xls). We get 0.00333, 0.00217, and 0.00283, respectively. The diffusion coefficient for H_2S into air at this temperature is approximately 0.032 cm²/s (see Table A.2).

The Schmidt number is 4.93. The three Sherwood numbers are: 112, 733, and 9533, respectively. The three mass transfer coefficients are: K=0.000720, 0.000469, and 0.000610 m/s, respectively. Considering the maximum molar concentration at saturation (0.1) and the molecular weight (34.1), the three rates of transfer are: 8.84, 5.76, and 7.49 kg/hr/m², respectively. The inputs and results are shown in the following figure:

	A	B	C	D
1	user inputs		calculations	
2	1.25	V [m/s]	4.0E+04	Re1
3	0.5	x1 [m]	4.0E+05	Re2
4	5	x2 [m]	4.0E+06	Re3
5	50	x3 [m]	0.00333	f1
6	properties of air		0.00217	f2
7	1.145	ρ [kg/m³]	0.00283	f3
8	0.00001805	μ [kg/m/s]	4.93	Sc=$\mu/\rho D$
9	properties of H_2S		112	Sh1
10	34.1	g/mole	733	Sh2
11	0.032	D [cm²/s]	9533	Sh3
12	0.1	conc.	0.000720	K1 [m/s]
13			0.000469	K2 [m/s]
14			0.000610	K3 [m/s]
15			8.84	J1 [kg/hr/m²]
16			5.76	J2 [kg/hr/m²]
17			7.49	J3 [kg/hr/m²]

Figure 28. Example 11 Results

Flow over a Cylinder

For our next example we will consider flow over a cylinder (rod, pipe, tube). This configuration is often used for industrial processes. Here we consider contamination from old pipes containing up to 2% mercury. Air flows across these pipes, picking up mercury and dispersing it into the atmosphere. We see from Table A1 that the diffusion coefficient for Hg into N_2 is 0.119 cm²/s (0.0000119 m²/s). Air is 79% nitrogen and so we will use this value in our calculations. The pipes are standard for the 1950s when this mess was assembled, nominally 1 inch, making the actual outside diameter 3.34 cm. This calculation was made to estimate the risk of airborne mercury after the pipes were unearthed and before they were reburied in a more suitable location.

Churchill and Bernstein[22] provide a very useful convective heat transfer correlation for flow over a cylinder that is accurate over a wide range of Reynolds numbers, ideal for this example.

[22] Churchill, S. W., and Bernstein, M., "A Correlating Equation for Forced Convection from Gases and Liquids to a Circular Cylinder in Crossflow", *Journal of Heat Transfer*, Vol. 99, No. 2, pp. 300–306, 1977.

$$Nu = 0.3 + \frac{0.62 Re^{\frac{1}{2}} Pr^{\frac{1}{3}}}{\left[1 + \left(\frac{0.4}{Pr}\right)^{\frac{2}{3}}\right]^{\frac{1}{4}}} \left[1 + \left(\frac{Re}{282,000}\right)^{\frac{5}{8}}\right]^{\frac{4}{5}}$$ (7.7)

While this empirical correlation is intended for heat transfer, we can use the Chilton-Colburn Analogy and adapt it to mass transfer. Combining Equations 6.2 and 6.11, we get:

$$Sh = \frac{Nu Sc^{\frac{1}{3}}}{Pr}$$ (7.8)

We perform these calculations in spreadsheet example12.xls. The density, ρ, is 1.184 kg/m³ and dynamic viscosity, μ, is 0.0186 cP (0.0000186 kg/m/s), making the Reynolds number at a velocity of 1 m/s:

$$Re = \frac{\rho V d}{\mu} = \frac{\left(1.184 \frac{kg}{m^3}\right)\left(1 \frac{m}{s}\right)(0.0334m)}{\left(0.0000186 \frac{kg}{ms}\right)} = 2126$$ (7.9)

The constant pressure specific heat, Cp, for air at 25°C is 1.005 kJ/kg/°C and the thermal conductivity, k, at this temperature is, 0.02624 W/m/°C, making the Prandtl number:

$$Pr = \frac{\mu C}{k} = \frac{\left(0.0000186 \frac{kg}{ms}\right)\left(1005 \frac{J}{kg°C}\right)}{\left(0.02624 \frac{W}{m°C}\right)} = 0.712$$ (7.10)

We consider air velocities of 0.5 to 3.0 m/s. The Nusselt number at 1.0 m/s is 26.2, which we use to calculate a Stanton number of 0.0173. The density of mercury vapor at such temperatures is exceedingly small: 0.000106 kg/m³. The viscosity would also be exceedingly small at this point, making the ratio of the two meaningless. Therefore, we must use the properties of air when calculating the Schmidt number, which is consistent with presuming the diffusion coefficient of Hg into air is the same as air into Hg:

$$Sc = \frac{\mu}{\rho D} = \frac{\left(0.0000186\,\frac{kg}{ms}\right)}{\left(1.184\,\frac{kg}{m^3}\right)\left(0.0000119\,\frac{m^2}{s}\right)} = 1.32 \tag{7.11}$$

Using Equation 7.8 and these results we find the Sherwood number at an air velocity of 1 m/s to be 40.4. Equation 7.4 then yields the mass transfer coefficient:

$$K = \frac{ShD}{d} = \frac{40.4\left(0.0000119\,\frac{m^2}{s}\right)}{(0.0334m)} = 0.0144\,\frac{m}{s} \tag{7.12}$$

If the source concentration were 2% and the density as stated above, this would indicate a flux of 0.0115 grams of Hg per hour per meter of exposed pipe. Even considering the number and length of pipe involved, the risk was considered acceptable for the limited duration of the remediation project. The results from example12.xls are listed below.

	A	B	C	D	E	F	G	H	I	J	K
1	user inputs		properties of Hg		calculations						
2	3.34 d [cm]		0.000106 ρ [kg/m³]		V [m/s]	Re	Nu	St	Sh	K [m/s]	J [g/hr/m]
3	properties of air		0.119 D [cm²/s]		0.5	1063	17.8	0.0235	27.4	0.0098	0.0078
4	1.184 ρ [kg/m³]		2% conc.		1.0	2126	26.2	0.0173	40.4	0.0144	0.0115
5	1.005 Cp [kJ/kg/°C]		calculations		1.5	3189	33.3	0.0146	51.2	0.0182	0.0146
6	0.0000186 μ [kg/m/s]		0.712 Pr		2.0	4252	39.6	0.0131	60.9	0.0217	0.0174
7	0.02624 k [W/m/°C]		1.32 Sc		2.5	5315	45.4	0.0120	70.0	0.0249	0.0200
8					3.0	6378	51.0	0.0112	78.5	0.0280	0.0224

Figure 29. Example 12 Results

Mass Transfer for Falling Droplets

Falling droplets are part of many industrial processes. One examples is evaporative cooling. Empirical correlations are readily available for this process, including that of Krieth and Black[23]:

$$Sh = 2 + \left(0.4Re^{\frac{1}{2}} + 0.6Re^{\frac{2}{3}}\right)Sc^{0.4} \tag{7.13}$$

We could also use the Chilton-Colburn Analogy, as drag (or friction factor) for a sphere is well known. Note that in this case $Cd=4f$, which can be tricky. The figure below is typical:

---

[23] Krieth, F. and W. Z. Black, *Basic Heat Transfer*, Harper and Rowe, 1980.

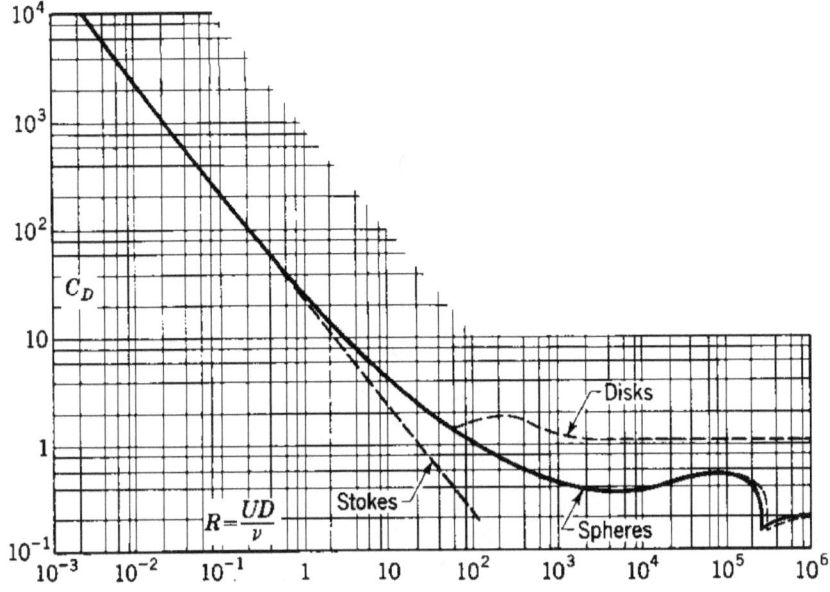

Figure 30. Drag Coefficient for a Sphere

We will use the same air properties as in example 12. The calculations and results may be found in example13.xls in the examples folder of the online archive accompanying this text. Experimental measurements indicate that water droplets used in cooling applications are usually between 1 and 5 mm with 3 mm being typical. While these vary widely, we will consider velocities of the air relative to the droplets ranging from 0.5 m/s to 5.0 m/s.

The density of saturated water vapor at 25°C is 0.0000185 kg/m³. The diffusion coefficient of water vapor into air at this same temperature is 0.3 cm²/s (see Table A2). The concentration of water at the interface with air is 100%. The Prandtl number is still 0.712 and the Schmidt number for this combination is 0.534. The user inputs for this problem are shown below:

	A	B
1	user inputs	
2	properties of air	
3	1.184	ρ [kg/m³]
4	1.005	Cp [kJ/kg/°C]
5	0.0000186	μ [kg/m/s]
6	properties of H_2O	
7	0.000106	ρ [kg/m³]
8	0.3	D [cm²/s]
9	100%	conc.
10	calculations	
11	0.524	Sc

Figure 31. Example 13 User Inputs

The Reynolds numbers for each diameter and velocity are shown in this next table:

E	F	G	H	I	J
Reynolds Number					
	diameter, d [mm]				
V [m/s]	1	2	3	4	5
0.5	32	64	95	127	159
1.0	64	127	191	255	318
1.5	95	191	286	382	477
2.0	127	255	382	509	637
2.5	159	318	477	637	796
3.0	191	382	573	764	955
3.5	223	446	668	891	1114
4.0	255	509	764	1018	1273
4.5	286	573	859	1146	1432
5.0	318	637	955	1273	1591

Figure 32. Example 13 Reynolds Numbers

These are the same regardless of the empirical correlation used. The first group of calculations based on Equation 7.13 are shown in the table on the next page.

Sherwood Number, Sh (Eqn. 7.13)

V [m/s]	diameter, d [mm]				
	1	2	3	4	5
0.5	8.4	11.8	14.7	17.2	19.5
1.0	11.8	17.2	21.6	25.5	29.1
1.5	14.7	21.6	27.4	32.4	37.0
2.0	17.2	25.5	32.4	38.5	44.1
2.5	19.5	29.1	37.0	44.1	50.5
3.0	21.6	32.4	41.3	49.2	56.5
3.5	23.6	35.5	45.4	54.1	62.1
4.0	25.5	38.5	49.2	58.7	67.4
4.5	27.4	41.3	52.9	63.2	72.5
5.0	29.1	44.1	56.5	67.4	77.5

mass transfer coefficient, K [m/s]

V [m/s]	diameter, d [mm]				
	1	2	3	4	5
0.5	0.252	0.178	0.147	0.129	0.117
1.0	0.355	0.258	0.216	0.192	0.175
1.5	0.441	0.324	0.274	0.243	0.222
2.0	0.516	0.383	0.324	0.289	0.264
2.5	0.585	0.437	0.370	0.331	0.303
3.0	0.649	0.486	0.413	0.369	0.339
3.5	0.709	0.533	0.454	0.406	0.372
4.0	0.766	0.578	0.492	0.441	0.405
4.5	0.821	0.620	0.529	0.474	0.435
5.0	0.873	0.661	0.565	0.506	0.465

evaporation [μg/s/droplet]

V [m/s]	diameter, d [mm]				
	1	2	3	4	5
0.5	0.084	0.237	0.440	0.688	0.974
1.0	0.118	0.344	0.648	1.020	1.454
1.5	0.147	0.432	0.820	1.295	1.850
2.0	0.172	0.510	0.972	1.539	2.201
2.5	0.195	0.581	1.110	1.761	2.522
3.0	0.216	0.648	1.239	1.968	2.820
3.5	0.236	0.710	1.360	2.162	3.101
4.0	0.255	0.769	1.476	2.347	3.368
4.5	0.273	0.826	1.586	2.524	3.624
5.0	0.291	0.880	1.692	2.695	3.869

Figure 33. Example 13 Calculations Using Equation 7.13

The drag coefficients are shown in this next table:

Drag Coefficient, Cd=4f				
diameter, d [mm]				
V [m/s]	1	2	3	4	5
0.5	1.984	1.350	1.101	0.961	0.870
1.0	1.350	0.961	0.805	0.717	0.659
1.5	1.101	0.805	0.685	0.617	0.572
2.0	0.961	0.717	0.617	0.560	0.522
2.5	0.870	0.659	0.572	0.522	0.489
3.0	0.805	0.617	0.539	0.495	0.466
3.5	0.756	0.585	0.515	0.474	0.447
4.0	0.717	0.560	0.495	0.458	0.433
4.5	0.685	0.539	0.479	0.444	0.422
5.0	0.659	0.522	0.466	0.433	0.412

Figure 34. Example 13 Drag Coefficients (Cd=4f)

The group of calculations using Equation 7.3 are listed in the table on the next page.

Figure 35. Streamlines around a Sphere without Separation

46

Sherwood Number, Sh (Eqn. 7.3)					
	diameter, d [mm]				
V [m/s]	1	2	3	4	5
0.5	6.4	8.7	10.6	12.3	14.0
1.0	8.7	12.3	15.5	18.4	21.1
1.5	10.6	15.5	19.8	23.7	27.5
2.0	12.3	18.4	23.7	28.7	33.5
2.5	14.0	21.1	27.5	33.5	39.2
3.0	15.5	23.7	31.1	38.1	44.8
3.5	17.0	26.3	34.7	42.6	50.2
4.0	18.4	28.7	38.1	47.0	55.6
4.5	19.8	31.1	41.5	51.3	60.8
5.0	21.1	33.5	44.8	55.6	66.1
mass transfer coefficient, K [m/s]					
	diameter, d [mm]				
V [m/s]	1	2	3	4	5
0.5	0.191	0.130	0.106	0.092	0.084
1.0	0.260	0.185	0.155	0.138	0.127
1.5	0.318	0.232	0.198	0.178	0.165
2.0	0.370	0.276	0.237	0.215	0.201
2.5	0.419	0.317	0.275	0.251	0.235
3.0	0.465	0.356	0.311	0.286	0.269
3.5	0.509	0.394	0.347	0.319	0.301
4.0	0.552	0.431	0.381	0.352	0.333
4.5	0.593	0.467	0.415	0.385	0.365
5.0	0.634	0.502	0.448	0.417	0.396
evaporation [µg/s/droplet]					
	diameter, d [mm]				
V [m/s]	1	2	3	4	5
0.5	0.064	0.173	0.317	0.493	0.697
1.0	0.086	0.246	0.464	0.735	1.055
1.5	0.106	0.310	0.593	0.949	1.374
2.0	0.123	0.368	0.712	1.148	1.673
2.5	0.139	0.422	0.824	1.338	1.959
3.0	0.155	0.474	0.933	1.522	2.237
3.5	0.170	0.525	1.039	1.701	2.509
4.0	0.184	0.574	1.142	1.877	2.776
4.5	0.198	0.622	1.243	2.050	3.039
5.0	0.211	0.669	1.342	2.221	3.299

Figure 36. Example 13 Calculations Using Equation 7.3

A comparison of Sherwood numbers calculated using Equation 7.13 and Equation 7.3 are shown in this next figure.

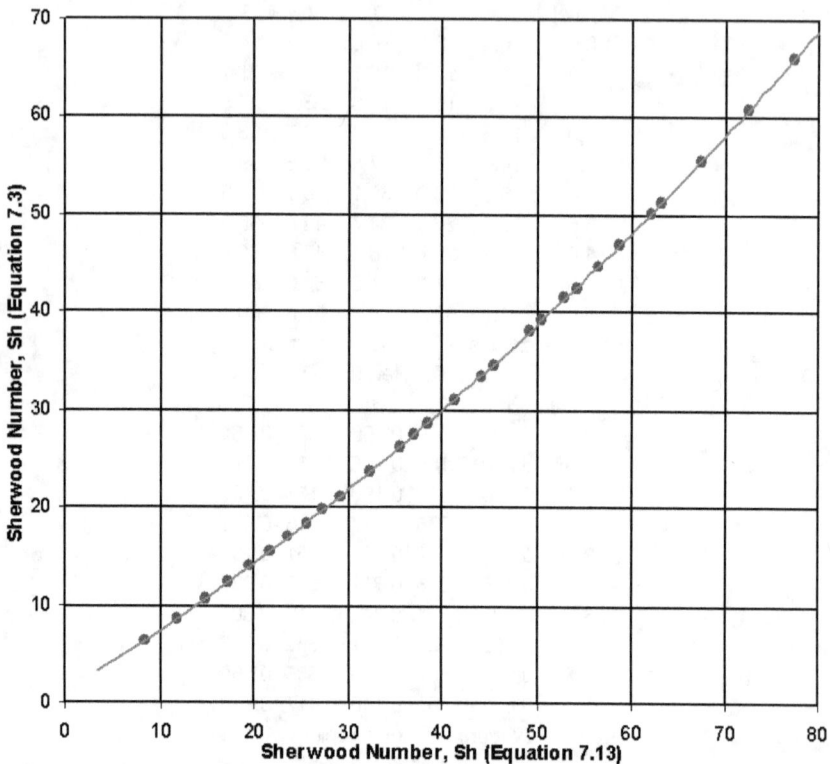

Figure 37. Example 13 Comparison of Sherwood Numbers

Discrepancies of this magnitude are not uncommon.

A comparison of mass flux in micro grams per second per droplet for the two different Sherwood numbers is shown in this next figure.

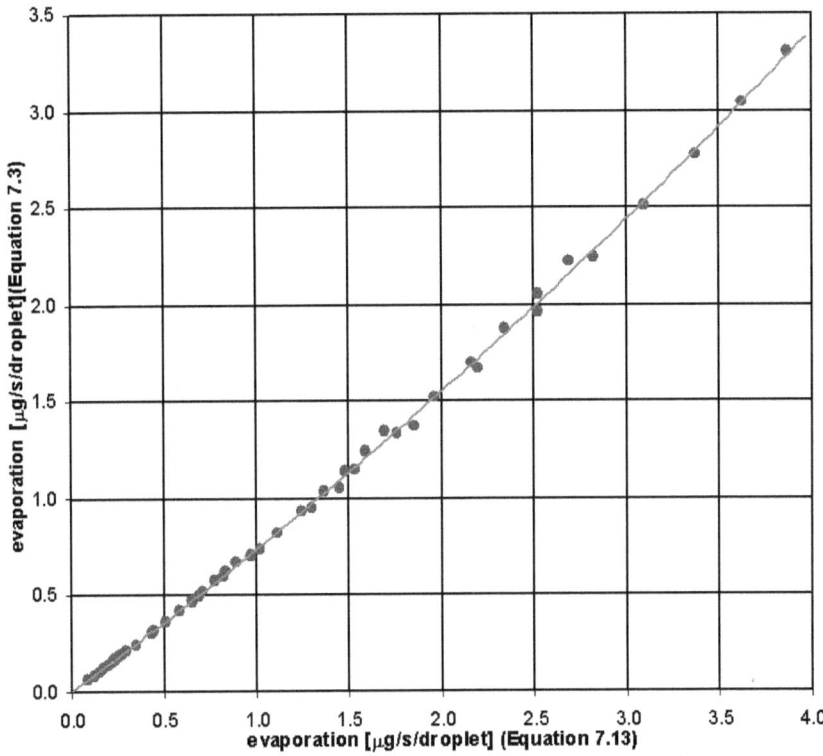

Figure 38. Example 13 Comparison of Mass Fluxes

Chapter 8. Sprays

Sprays are perhaps the most efficient industrial mass transfer process. In the previous chapter we considered mass transfer for a single droplet. While this is an interesting problem, it has only limited use. Most processes involve a spray or a volume filled with numerous droplets. While we could simply multiply the results for one droplet, this is not the most efficient or accurate way of handling the matter. The preferred way is to take measurements, develop, and utilize empirical correlations for the spray in question. We begin by considering a typical spray, such as that illustrated in this next figure:

Figure 39. Water Spray with Typical Droplets

Some sort of nozzle is often used to create a spray, as illustrated in the figure above. The water droplets in this figure are typical in size, approximately 2-3 mm in diameter.

Even with the same nozzle, changing the pressure can change the size of the droplets, as illustrated in this next figure:

Figure 40. Water Spray with Much Smaller Droplets

The droplets are now approximately 0.5-1.0 mm in diameter. Any empirical correlation must consider the size of droplets or be limited to a particular spray nozzle and pressure combination yielding a range of droplet sizes. When applying an empirical correlation, make sure your application is comparable at this level. While it may be tempting to simply adjust a spray correlation based on differing droplet sizes, like those calculated in the preceding chapter, this may not yield adequately accurate results, as the interaction between droplets is not insignificant.

Combined Heat and Mass Transfer

One example of this type of spray occurs inside an evaporative cooling tower. Droplets are created by water flowing through specially designed nozzles. The droplets fall for some distance, sometimes encountering packing, which we discuss in the next chapter. When the air flows vertically upward through the packing and the water vertically downward, this arrangement is called *counterflow*. This is distinguished from designs where the air flows sideways through the packing as the water falls vertically downward, which is called *crossflow*. Below the packing, water droplets fall vertically downward and air flows in from the sides and then up through the center so that this *rain zone* is not strictly crossflow or counterflow. Still, it is a common arrangement in many applications so that calculations have been made, experiments have been conducted, and correlations have been developed.

The interaction between falling droplets of water and air encompasses both heat and mass transfer. To distinguish between these two, we often add the qualifiers: *sensible* heat transfer and *evaporative* heat (or mass) transfer. There is no way to separate the two in practice. Rather than try to separate them, we purposefully combine the two processes.

Merkel[24] proposed a theory relating the evaporation and sensible heat transfer (occurring in a direct contact process such as a spray) to the enthalpy difference. Merkel made six basic assumptions, introduced at various points in the development to simplify the mathematics. There are many implications arising from these assumptions beyond the scope of this text. More details can be found in a paper written by the author with Al Feltzin published as a paper and also in the Journal of the Cooling Technology Institute.[25]

The model on which Merkel's theory was developed consists of a water droplet at temperature, T, surrounded by a thin air film (interface). Merkel assumed that the air film is saturated and therefore is also at temperature T_A. Thus the film has humidity ratio, W_F, and enthalpy, h_F. Surrounding the air film is the bulk air mass at some lower temperature $T_A < T_F$, and humidity ratio, $W_A < W_F$, and an enthalpy $h_A < h_F$.

---

[24] Merkel, F. Verdunstungskulung, V.D.I. Forschungsarbeiteh, No. 275, Berlin, 1925.

[25] Feltzin, A. E., and Benton, D. J., "A More Nearly Exact Representation of Cooling Tower Theory," CTI Technical Paper No. TP91-02.

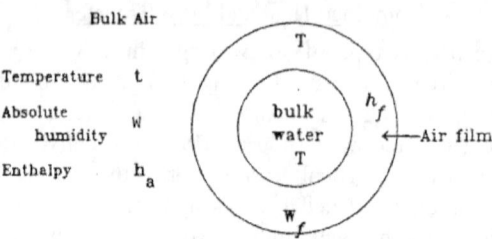

Bulk Air

Temperature   t

Absolute
humidity   W

Enthalpy h_a

Figure 41. Merkel's Droplet Model

If a is the interfacial surface (length²/length³) and V is the contacting volume (length³) then the interfacial surface area, $S=aV$ (length) and the differential surface of the model droplet interfacial film is $dS=adV$ (surface element). The flux of water is given the symbol, L (mass/time/length²) and the flux of air, G (lbs/hr/ft²). The two fluxes are opposing, that is, *counterflow*. The preceding figure illustrates this process and the variables.

Heat is transferred from the water droplet to the bulk air through the interface by two means, sensible heat transfer (convection, due to a difference in temperature) and latent heat of evaporation (mass transfer by diffusion, due to a difference in concentration). Merkel assumed that the interface offers no resistance to heat transfer from the water droplet to the bulk air by either of these mechanisms. The sensible heat transfer rate by convection is given by:

$$dq_S = K_C \left(T_F - T_A \right) adV \tag{8.1}$$

where K_C is the convective heat transfer coefficient, energy/time/length²/temp. The mass transfer rate is given by:

$$dL = K_M \left(P_F - P_A \right) adV \tag{8.2}$$

where K_M is a diffusional mass transfer coefficient. P_F and P_A are the partial pressures of water vapor in the interfacial film at temperature, T_F, and bulk air at temperature, T_A, respectively. Next, Merkel assumed that the partial pressure of water vapor is proportional to humidity, that is:

$$P_F \propto W_F$$
$$P_A \propto W_A \tag{8.3}$$

which can be substituted into Equation 8.2 to obtain:

$$dL = K_M \left(W_F - W_A \right) adV \tag{8.4}$$

The evaporative (latent) heat transfer rate due to diffusional mass transfer is given by:

$$dq_L = \lambda dL = \lambda K_M \left(W_F - W_A \right) adV \tag{8.5}$$

where λ is the latent heat of vaporization. The total transfer rate is then given by:

$$dq_{total} = \left[K_C \left(T_F - T_A \right) + \lambda K_M \left(W_F - W_A \right) adV \right] \qquad (8.6)$$

At this point in the derivation, the concept of humid heat, C_S, (the heat capacity of an air-water vapor mixture) is usually introduced. By addition and subtraction of a term $C_S(T_W\text{-}T_A)$ to the right hand side of Equation 8.6 and algebraic manipulation, we arrive at the following:

$$dq_{total} = K_M \left\{ \begin{array}{c} \left(C_S T_F + \lambda W_F \right) - \left(C_S T_A + \lambda W_A \right) \\ + C_S \left(T_F - T_A \right) \left[\dfrac{K_C - 1}{C_S K_M} \right] \end{array} \right\} adV \qquad (8.7)$$

Merkel also assumed that the Lewis number, L_E, of unity, that is:

$$L_E = \frac{Sc}{Pr} = \frac{K_C}{C_P K_M} = 1 \qquad (8.8)$$

This assumption causes the last term in Equation 8.7 to vanish. The terms $C_S T_F + \lambda W_F$ and $C_S T_A + \lambda W_A$ are close to, but not exactly, reducing the heat transfer to:

$$dq = K \left(h_S - h_A \right) adV \qquad (8.9)$$

The subscript on K has been dropped, as there is now only one. The subscript F (film) has been replaced with S (saturation), in keeping with the previous assumption. Conservation of energy requires that:

$$dq = d \left(L C_{PW} T_W \right) = d \left(G h_A \right) \qquad (8.10)$$

where C_{PW} is the constant pressure specific heat of water in the liquid state. These two equations can be combined to form:

$$dq = C_{PW} \left(L dT_W - T_W dL \right) = G dh_A = K \left(h_F - h_A \right) adV \qquad (8.11)$$

Dividing by $h_F\text{-}h_A$ and integrating yields:

$$KaV = C_{PW} \left\{ \int_{T_{OUT}}^{T_{IN}} \frac{L dT}{\left(h_F - h_A \right)} + \int_{T_{OUT}}^{T_{IN}} \frac{T dL}{\left(h_F - h_A \right)} \right\} \qquad (8.12)$$

$$KaV = G \int_{h_{ain}}^{h_{aout}} \frac{dh_A}{\left(h_F - h_A \right)} \qquad (8.13)$$

Merkel went on to assume that the portion of the water evaporated was insignificant to the whole, or $dL=0$. After making this assumption, Equations 8.12 and 8.13 become:

$$\frac{KaV}{L} = C_{PW} \int_{T_{OUT}}^{T_{IN}} \frac{dT}{h_F - h_A} = \frac{G}{L} \int_{h_{ain}}^{h_{aout}} \frac{dh}{\left(h_F - h_A\right)} \qquad (8.14)$$

The third term in Equation 8.14 is of no further interest at this point. This leaves the classic form of Merkel's Equation:

$$\frac{KaV}{L} = C_{PW} \int_{T_{OUT}}^{T_{IN}} \frac{dT}{h_F - h_A} \qquad (8.15)$$

This is a nonlinear integral equation with no analytical solution and must be solved numerically. Merkel chose to use the 4-point Chebyshev method for its simplicity. In spite of these assumptions and simplifications, Merkel's Equation served the cooling tower industry well for decades. See Appendix C for details.

Supply and Demand Curves

Even with all of the preceding theoretical development, there remains the question of how to utilize this information? This leads to the concept of supply and demand. Cooling some rate of flowing water (here in the form of falling droplets) from one temperature to another requires a corresponding rate of combined sensible heat and evaporative heat (and/or mass) transfer. This is the *demand*. A volume of such falling droplets swept by an upward flowing stream of air provides a certain capacity for such transfer, which we call the *supply*.

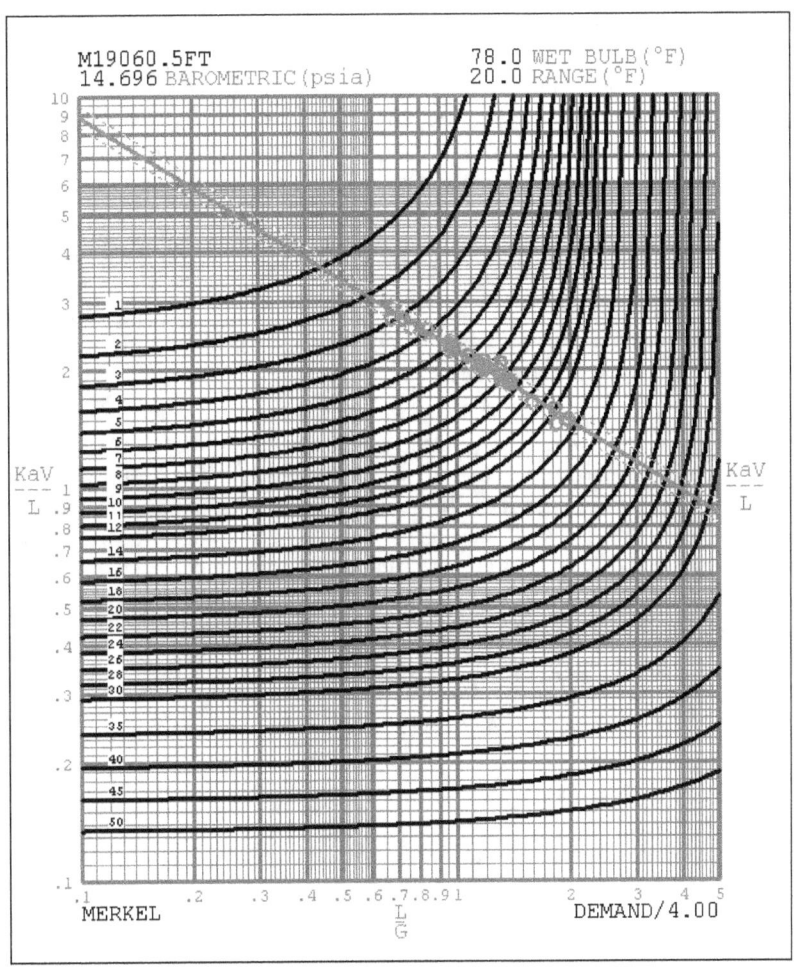

Figure 42. Typical Merkel Demand Curves

Designing a system or calculating the performance of an existing system then requires matching *supply* and *demand*. We calculate the demand using Merkel's theory (Equation 8.15, Appendix C) to obtain the black curves in the preceding figure. We conduct experiments and develop correlations to quantify the supply (or capacity) of various sprays to obtain the red points and red regression line in the preceding figure. We note here without elaboration that approximately 80% of the total is due to evaporation, while 20% is due to sensible (i.e., convection). This follows from presuming the Lewis number is unity. More details on this may be found in the paper cited previously and in Appendix D.

The Cooling Technology Institute[26] using Equation 8.15 and the 4-point Chebyshev method of numerical integration, expanded on earlier KaV/L vs. L/G demand curves, such as those utilized by Foster Wheeler Corporation[27] and J.F. Pritchard Company.[28] The CTI curves had the advantages of being computer generated and computer drawn, and made what had been very limited published data much more widely available and over a much broader selection of ranges and approaches. These curves served the industry for decades.

These *demand* curves were used to obtain graphical solutions of cooling tower performance, which was a tedious process. The loose-leaf notebook filled with such curves was an expensive item and hard to come by. The same curves can now easily be generated with an Excel® spreadsheet, as illustrated in the previous figure and spreadsheet Merkel.xls.

In these figures, the black curves are constant *approach* (cold water temperature minus ambient wet-bulb). The *range* is the hot (entering) water temperature minus the cold (leaving) water temperature. The horizontal axis is the ratio of the water to air mass flows. The downward-sloping red line in the previous figure is the *supply* line, or the dimensionless mass transfer capacity provided by the packing material, in this case water droplets.

Lowe & Christie[29] provide several curves for falling droplets. These have been combined to yield a generalized correlation over a range of droplet sizes. While this correlation does not cover all sprays, it has been compared to experimental data[30] and shown to be adequate. These spray *supply* curves can be found in spreadsheet spray.xls and are shown in this next figure:

---

[26] *CTI Blue Book of Counterflow Demand Cooling Curves*, CTI, Houston, Texas, 1967.

[27] *Cooling Tower Performance: Bulletin CT432*, Foster Wheeler Corporation, 1943.

[28] *Counterflow Cooling Tower Performance*, F. Pritchard Company of California, 1957.

[29] Lowe, H. J., and D. G. Christie, "Heat Transfer and Pressure Drop Data on Cooling Tower Packings and Model Studies of the Resistance of Natural Draught Towers to Airflow," International Division of Heat Transfer, Part V, pp. 333-950, American Society of Mechanical Engineers, New York, 1961.

[30] Benton, D. J. and R. L. Rehberg, "Mass Transfer and Pressure Drop in Sprays Falling in a Freestream at Various Angles," International Association for Hydraulic Research, Fifth Cooling Tower Workshop, Palo Alto, California, September 29 - October 3, 1986.

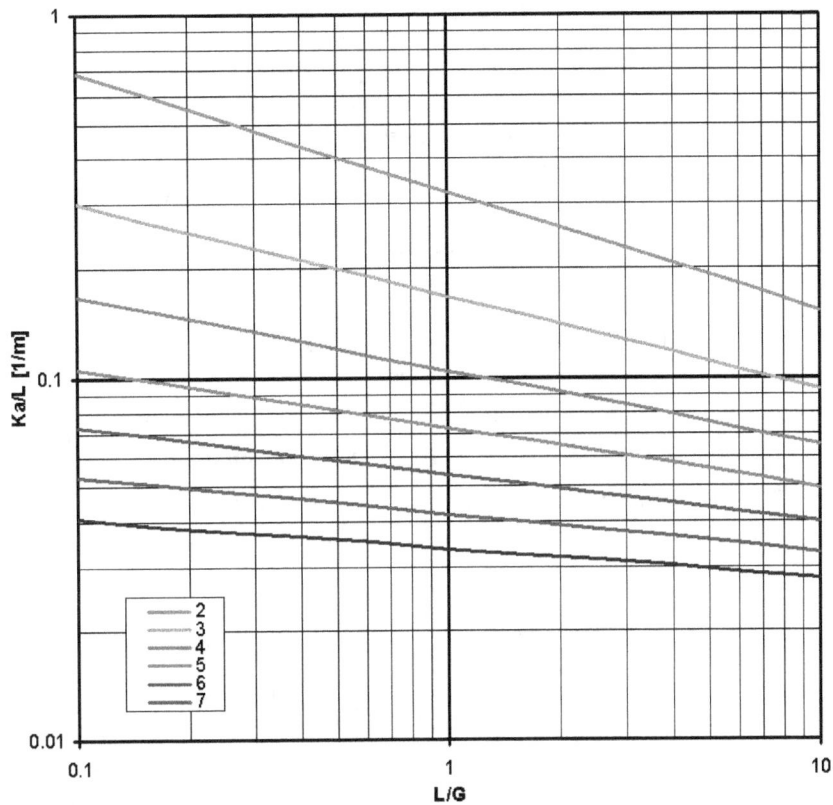

Figure 43. Supply Curves for Falling Drops

The horizontal axis is the ratio of water to air flows (or fluxes) and the vertical axis is the mass transfer factor, *Ka/L*, per meter of depth. The correlation is calculated with the following Excel macro:

```
Function spray(LG As Double, d As Double) As Double
   Dim x As Double, y As Double
   x=log10(LG)
   y=log10(d)
   spray=10^(-0.449177*x-1.6298*y+0.40809*x*y)
End Function
```

We can combine the previous Merkel *demand* with the spray *supply* to see what cooling can be expected for a particular depth and droplet diameter. These calculations can be found on Sheet2 of the spreadsheet spray.xls.

Typical calculations are shown in the figure below:

	A	B
1		spray cooling - water droplets
2		user inputs
3	2.5	droplet diameter, d [mm]
4	1.5	spray height, Y [m]
5	40	inlet water temperature [°C]
6	15	inlet wet-bulb temperature [°C]
7	1.25	flux ratio, L/G [unitless]
8		calculations
9	0.316	KaY/L [unitless]
10	33.7	exit water temperature [°C]
11	6.3	cooling range [°C]
12	18.7	approach [°C]

Figure 44. Typical Spray Cooling Calculations

Typical cold water temperature (Tcw) curves are shown in this next figure:

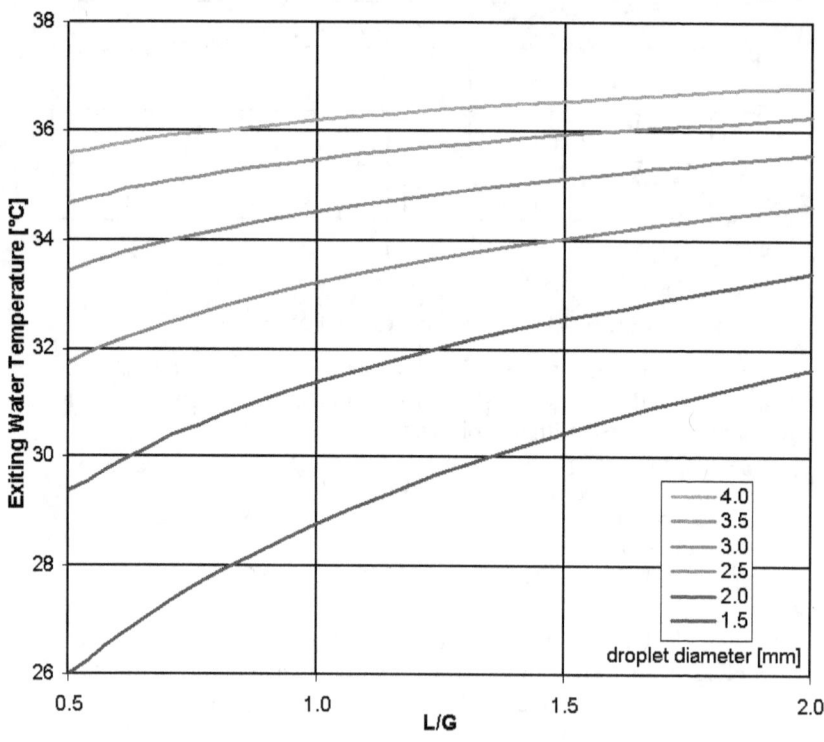

Figure 45. Impact of L/G and Droplet Diameter on Tcw

Matching the supply and demand becomes a nonlinear equation, which must be solved. This is accomplished by a macro within spray.xls using the bisection search method, a quick and simple approach that will always produce a result and never fails to converge. While there are methods that converge more rapidly, these may fail to converge under certain conditions. This consideration plus the trivial time savings makes the bisection search the natural choice.

The preceding figure showed the impact of droplet size on exiting water temperature vs. water to air flow rates, **L/G**. This next figure shows the impact of spray height.

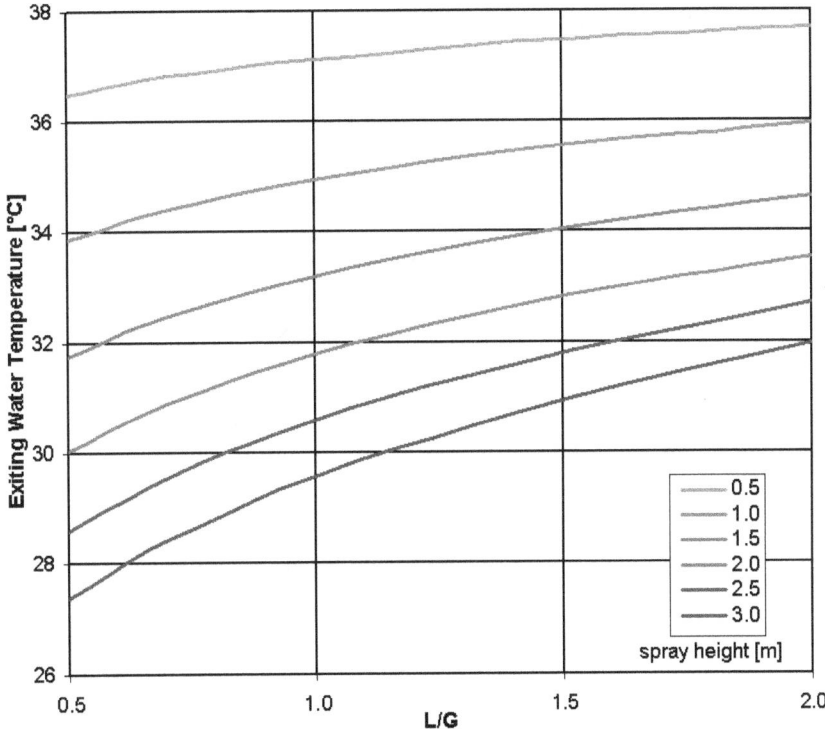

Figure 46. Impact of L/G and Spray Height on Tcw

While this is not necessarily true for all mass transfer or cooling processes, it does appear from these two figures that the impact of droplet size and spray height are similar in magnitude plus the curves have much the same shape. This means that the same result can be obtained by adjusting droplet size and/or spray height. Differences between applications will drive the decision as to what combination of droplet diameter and spray height is most cost effective.

Chapter 9. Packing

Packing of various types and shapes are perhaps the most common industrial mass transfer media. Most often there are two streams, which must react in order to produce the desired product. Typically, the heavier stream (for example, water) flows down and the lighter stream (for example, air) flows up. The tank or shell, which forms the reactor, is filled with *packing*. This packing may be anything from sand to glass spheres to metal turnings from a lathe. The packing material may be metallic, plastic, glass, silicon dioxide (SiO_2 or sand), or even wood chips. In most applications, the packing material does not itself participate in the reaction, only facilitating it. This hopefully results in longevity, necessitating infrequent replacement. The two primary functions of packing are: 1) increase surface area and 2) increase contact time.

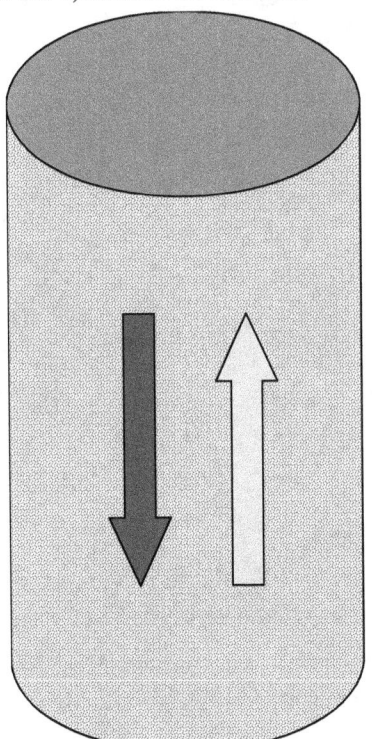

Figure 47. Packing Concept

The packing may be some natural substance (e.g., sand), specifically manufactured for this purpose (plastic spheres or honeycomb), or a waste product from some other process (e.g., metal turnings). Shapes and materials abound. Two types of plastic fill (spheres and rods) are shown in this next figure:

Figure 48. Typical Manufactured Plastic Packing Material

This packing is essentially a random and is simply dumped into the reactor, while other plastic packing may be formed into shapes.

This next type of plastic packing is very open and provides little surface area. Its main function is to increase contact time.

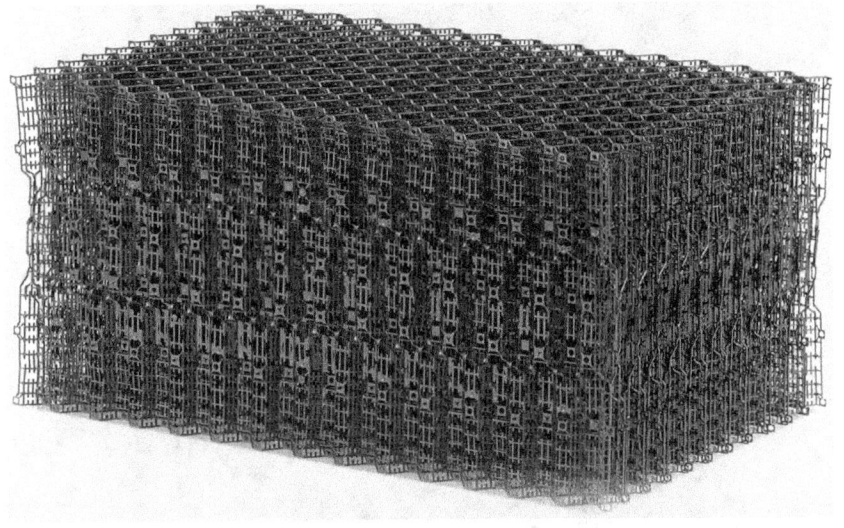

Figure 49. Open Plastic Packing

This next type of plastic packing increases surface area and contact time, as the water wets and flows down over the fluted surface:

Figure 50. Closed Plastic Packing

These last two types are used in cooling towers. Metal mesh, such as that shown below, is often used if erosion is a consideration.

Figure 51. Sturdy Regular Metallic Mesh

A mesh may also be somewhat randomly oriented or even consist of cut pieces, as illustrated in the picture below:

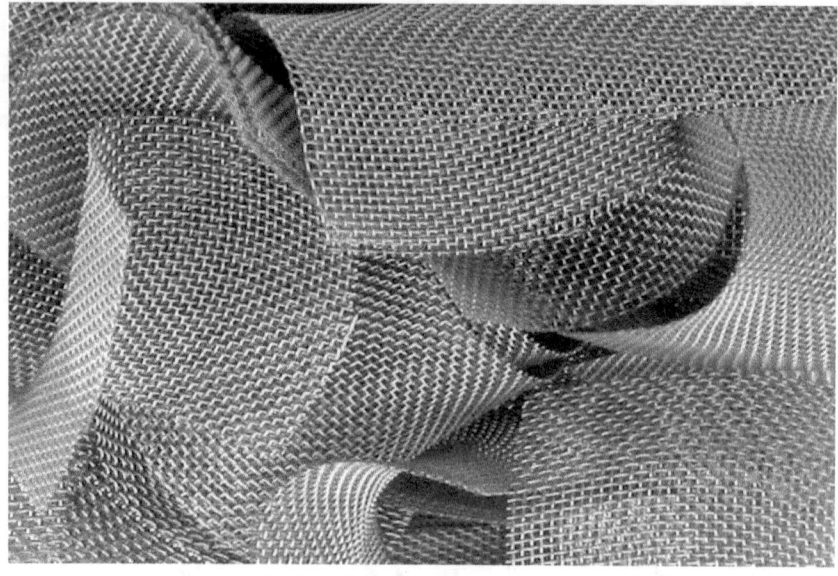

Figure 52. Randomly Oriented Metallic Mesh

The same calculations used for sprays can be applied to water falling through one of these types of packing (especially Figures 48 and 49) as air is forced upward, as in all counterflow cooling towers. In fact, the red circles in Figure 41 are for 5 feet (1.524 m) of Munters[31] type 19060 fill, very similar to that pictured in Figure 49. We can use the information in Figure 41 to create the following table of temperatures and required air flow, all at an ambient wet-bulb temperature of 78°F (25.55°C)

Table 1. 5ft / M19060 / 1 atm. / 78°F / 20°F

L/G	KaV/L	approach	Tcw	Thw
[]	[]	[°F]	[°F]	[°F]
0.44	3.68	1	79	99
0.58	3.10	2	80	100
0.69	2.70	3	81	101
0.84	2.48	4	82	102
0.96	2.31	5	83	103
1.08	2.14	6	84	104
1.21	2.02	7	85	105
1.34	1.91	8	86	106
1.44	1.81	9	87	107
1.57	1.72	10	88	108
1.69	1.64	11	89	109
1.81	1.55	12	90	110
2.08	1.44	14	92	112
2.37	1.34	16	94	114
2.64	1.26	18	96	116
2.91	1.17	20	98	118
3.29	1.11	22	100	120
3.60	1.04	24	102	122
3.96	0.99	26	104	124
4.34	0.93	28	106	126
4.63	0.89	30	108	128

The approach is the difference between the exiting water temperature and entering air temperature. Tcw and Thw are the cold (exiting) and hot (entering) water temperatures, respectively. Note that L/G and KaV/L are both unitless, hence the empty brackets [] beneath. These results along with the figure may be found in spreadsheet M19060.xls.

---

[31] The Munters Corporation manufacturers a variety of items, including sever types of packing material for cooling towers. They provide performance curves for their products on their website https://www.munters.com/en/

Surface Area

The increased surface area afforded by packing can be significant. Many correlations are available. We will only consider one here: hexagonal close packed. Unlike grains of sand, spheres of uniform size will tend toward a certain packing arrangement, as illustrated in this next figure:

Figure 53. Hexagonal Close Packed Spheres of Uniform Size

If the spheres are not of uniform size, other arrangements are possible:

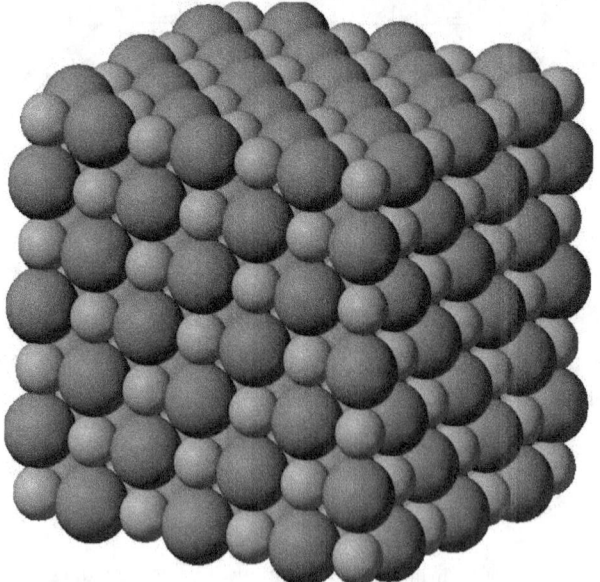

Figure 54. Packed Spheres of Non-Uniform Size

The maximum volumetric packing factor for spheres of uniform diameter was first proposed by Gauss to be:

$$\frac{V_{spheres}}{V_{total}} = \frac{\pi}{3\sqrt{2}} \approx 0.74048 \qquad (9.1)$$

As the spheres used in these applications are much smaller than the tank or reactor, we can use this relationship. Taking the formulas for the volume of a sphere ($V=\pi d^3/6$) and that of a rectangle ($V=hwd$) or cylinder ($V=\pi d^2h$) and the packing ratio above we can arrive at a formula for the approximate number of hexagonal close packed spheres that will fit inside a given volume:

$$n = \frac{\sqrt{2}V}{d^3} \qquad (9.2)$$

Multiplying by the surface area for a single sphere ($A=\pi d^2$) yields the total interfacial area:

$$A = \frac{\pi\sqrt{2}V}{d} \qquad (9.3)$$

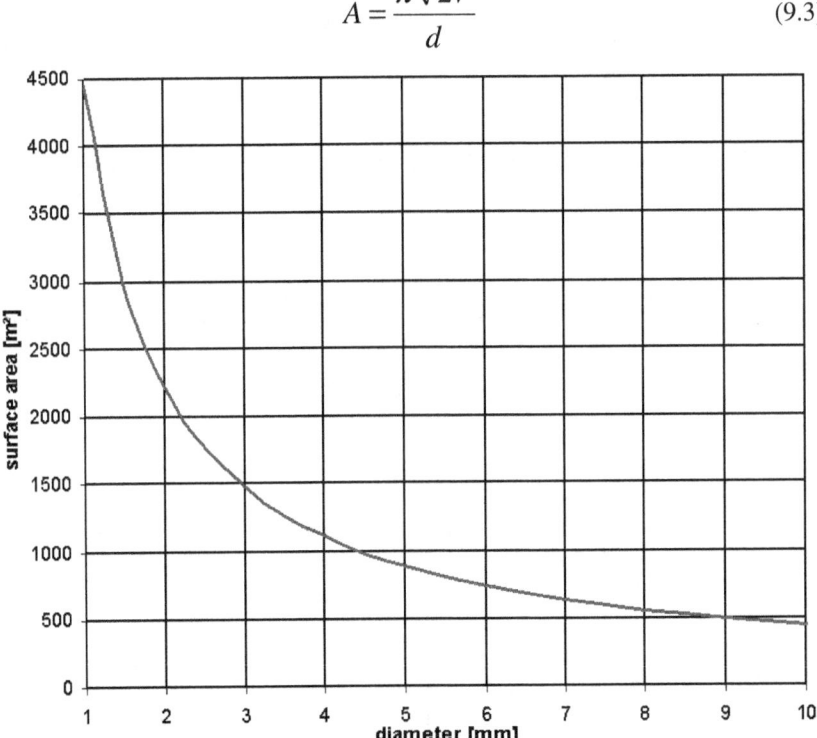

Figure 55. Surface Area vs. Packing Sphere Diameter

69

Packed Bed Example – Water and Methyl-Bromide

A bed packed with non-reactive spheres of uniform diameter (2 mm) is used to process Methyl-Bromide (CH_3Br). Water flows through the packing at a temperature of 35°C. The density of water at this temperature is 0.994 g/cm³ and the dynamic viscosity is 0.7191 cP (mPa-s). If the average velocity of water through the interstitial space between the spheres were 0.1 m/s, the Reynolds number would be:

$$Re = \frac{\rho V d}{\mu} = \frac{\left(994\frac{kg}{m^3}\right)\left(0.1\frac{m}{s}\right)(0.002m)}{\left(0.0007191\frac{kg}{ms}\right)} = 276 \qquad (9.4)$$

The diffusion coefficient is 1.43 cm²/g (Table A3), making the Schmidt number:

$$Sc = \frac{\mu}{\rho D} = \frac{\left(0.0007191\frac{kg}{ms}\right)}{\left(994\frac{kg}{m^3}\right)\left(0.000143\frac{m^2}{s}\right)} = 0.00506 \qquad (9.5)$$

Using Equation 7.13 for the Sherwood number yields: Sh=5.88. The mass transfer coefficient is then:

$$K = \frac{ShD}{d} = \frac{0.00506\left(0.000143\frac{m^2}{s}\right)}{(0.002m)} = 0.42\frac{m}{s} \qquad (9.6)$$

Using Equation 9.3 to calculate the surface area per cubic meter of reactor:

$$A = \frac{\pi\sqrt{2}\left(1m^3\right)}{0.002m} = 221\frac{m^2}{m^3} \qquad (9.7)$$

The concentration of Methly-Bromide is approximately 3%, making the flux per cubic meter of reactor:

$$J = KAC_0 = 28\frac{moles}{m^3 s} \qquad (9.8)$$

With a molecular weight of 94.94, this would be 2658 g/m³/s. As the Reynolds number is proportional to the velocity and the Sherwood number is proportional to the Reynolds number to a power of approximately 0.45 at this level, if we were to double the water velocity the flux would only increase by

36%. If we were to half the water velocity the flux would decrease by 23% to 77% of that at 0.1 m/s. If the spheres were 3 mm in diameter instead of 2 mm, while holding the velocity constant, the Reynolds number would increase and so would the Sherwood number, but the area would decrease, resulting in an overall decrease in flux of 47% to 53% of that for 2 mm spheres. If the spheres were 1 mm in diameter instead of 2 mm, holding the velocity constant, the flux would increase to 300%.

From these calculations we see that the performance of such a reactor would vary widely depending on the velocity and packing size, even if all of the other factors remained constant. There is also the pressure drop and operational stability to consider—both beyond the scope of this text. Clearly, designing an effective and efficient system requires careful consideration and experimental results, making experience in these matters of great value.

Liquid-Side Resistance in Gas Absorption

Vivian and Peaceman[32] studied absorption of CO_2 in columns characterized by wetted walls. After noting significant discrepancies between the prevailing quiescent surface theory, they embraced Higbie's penetration theory[33], which pictures the liquid as flowing over one piece of packing for a relatively short time before being mixed as it moves on to the next piece. In this model, absorption occurs during a series of brief contacts with a disrupted surface, rather than prolonged contact with a quiescent one. They provide the following empirical relationship:

$$Sh = 0.433 Sc^{\frac{1}{2}} Ga^{\frac{1}{6}} Re^{\frac{2}{5}} \qquad (9.9)$$

where **Ga** is the Galilei number:

$$Ga = \frac{\rho^2 g h^3}{\mu^2} \qquad (9.10)$$

The Galilei number is the ratio of the gravity forces to the viscous forces. This dimensionless number is often used when characterizing fluid films and liquids flowing down walls and over tubes, as in a steam surface condenser. The expectation is that this ratio will quantify the stability of the surface, as gravity tends to draw the liquid downward and viscosity tends to impede this movement in the form of drag on the wetted surface. Note that the Galilei number is often quite large, which is why the exponent is 1/6th in Equation 9.9.

---

[32] Vivian, J. E., and Peaceman, D. W., "Liquid-Side Resistance in Gas Absorption," AIChE Journal, 1956.

[33] Higbie, R., "Rate of Absorption of a Gas into a Still Liquid during Short Periods of Exposure," Transactions of the AIChE, Vol. 31, 1935.

Consider the case where CO_2 is in contact with water in a reactor, which consists primarily of flat, though not continuous, vertical surfaces. The temperature is 20°C and there is no other gas (e.g., air or N_2) in the system, making the concentration of CO_2 in the gas phase 100%. The diffusion coefficient for this pair at this condition is 1.67 cm²/s (Table A3). The density of water is 998.2 kg/m³ and the viscosity is 1.002 cP.

We will investigate a range of liquid velocities (0.01 to 0.1 m/s) and a range of wetted length (0.01 to 0.1 m). The calculations may be found in spreadsheet example14.xls.

	A	B
1	user inputs	
2	9.8	g [m²/s]
3	properties of H₂O	
4	998.2	ρ [kg/m³]
5	0.001002	μ [kg/m/s]
6	properties of CO₂	
7	100%	conc.
8	1.67	D [cm²/s]
9	1.84	ρ [kg/m³]
10	calculations	
11	0.00601	Sc

Figure 56. Example 14 User Inputs

	Reynolds Number, Re				
	length, l [m]				
V [m/s]	0.01	0.025	0.05	0.075	0.1
0.01	100	249	498	747	996
0.02	199	498	996	1494	1992
0.03	299	747	1494	2241	2989
0.04	398	996	1992	2989	3985
0.05	498	1245	2491	3736	4981
0.06	598	1494	2989	4483	5977
0.07	697	1743	3487	5230	6973
0.08	797	1992	3985	5977	7970
0.09	897	2241	4483	6724	8966
0.10	996	2491	4981	7472	9962

Figure 57. Example 14 Reynolds Numbers

	Galilei Number, Ga^(1/6)				
	length, l [m]				
V [m/s]	0.01	0.025	0.05	0.075	0.1
0.01	14.6	23.1	32.7	40.0	46.2
0.02	14.6	23.1	32.7	40.0	46.2
0.03	14.6	23.1	32.7	40.0	46.2
0.04	14.6	23.1	32.7	40.0	46.2
0.05	14.6	23.1	32.7	40.0	46.2
0.06	14.6	23.1	32.7	40.0	46.2
0.07	14.6	23.1	32.7	40.0	46.2
0.08	14.6	23.1	32.7	40.0	46.2
0.09	14.6	23.1	32.7	40.0	46.2
0.10	14.6	23.1	32.7	40.0	46.2

Figure 58. Example 14 Galilei Numbers (1/6th power)

	Sherwood Number, Sh				
	length, l [m]				
V [m/s]	0.01	0.025	0.05	0.075	0.1
0.01	3.1	7.0	13.2	18.9	24.5
0.02	4.1	9.3	17.4	25.0	32.4
0.03	4.8	10.9	20.4	29.4	38.1
0.04	5.4	12.3	22.9	33.0	42.7
0.05	5.9	13.4	25.0	36.1	46.7
0.06	6.3	14.4	26.9	38.8	50.3
0.07	6.7	15.4	28.6	41.3	53.5
0.08	7.1	16.2	30.2	43.5	56.4
0.09	7.4	17.0	31.7	45.6	59.1
0.10	7.8	17.7	33.0	47.6	61.7

Figure 59. Example 14 Sherwood Numbers

	mass transfer coefficient, K [m/s]				
	length, l [m]				
V [m/s]	0.01	0.025	0.05	0.075	0.1
0.01	516	471	439	422	410
0.02	681	621	580	557	541
0.03	801	731	682	655	636
0.04	898	820	765	734	714
0.05	982	896	836	803	780
0.06	1057	964	900	864	839
0.07	1124	1025	957	919	893
0.08	1185	1082	1009	969	942
0.09	1243	1134	1058	1016	987
0.10	1296	1183	1103	1060	1030

Figure 60. Example 14 Mass Transfer Coefficients

73

absorption [g/m²/s]				
length, l [m]				
V [m/s]	0.01	0.025	0.05	0.075	0.1
0.01	9.5	21.7	40.4	58.2	75.4
0.02	12.5	28.6	53.3	76.8	99.5
0.03	14.7	33.6	62.7	90.3	117.0
0.04	16.5	37.7	70.3	101.3	131.3
0.05	18.1	41.2	76.9	110.8	143.5
0.06	19.4	44.3	82.7	119.2	154.4
0.07	20.7	47.2	88.0	126.7	164.2
0.08	21.8	49.7	92.8	133.7	173.2
0.09	22.9	52.1	97.3	140.1	181.6
0.10	23.8	54.4	101.5	146.2	189.4

Figure 61. Example 14 Absorption Rates

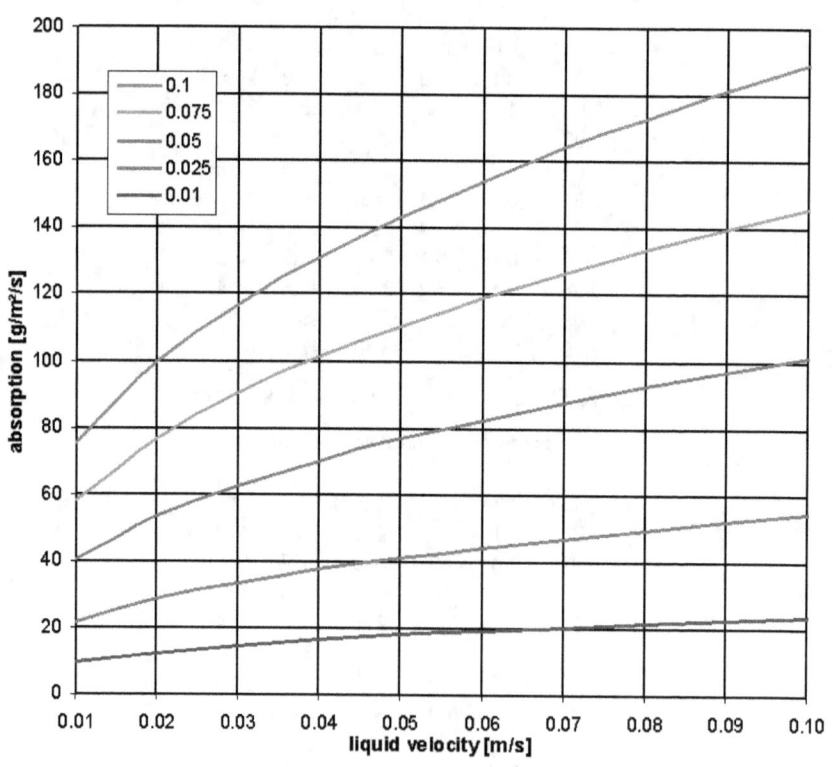

Figure 62. Example 14 Absorption Rate vs. Liquid Velocity

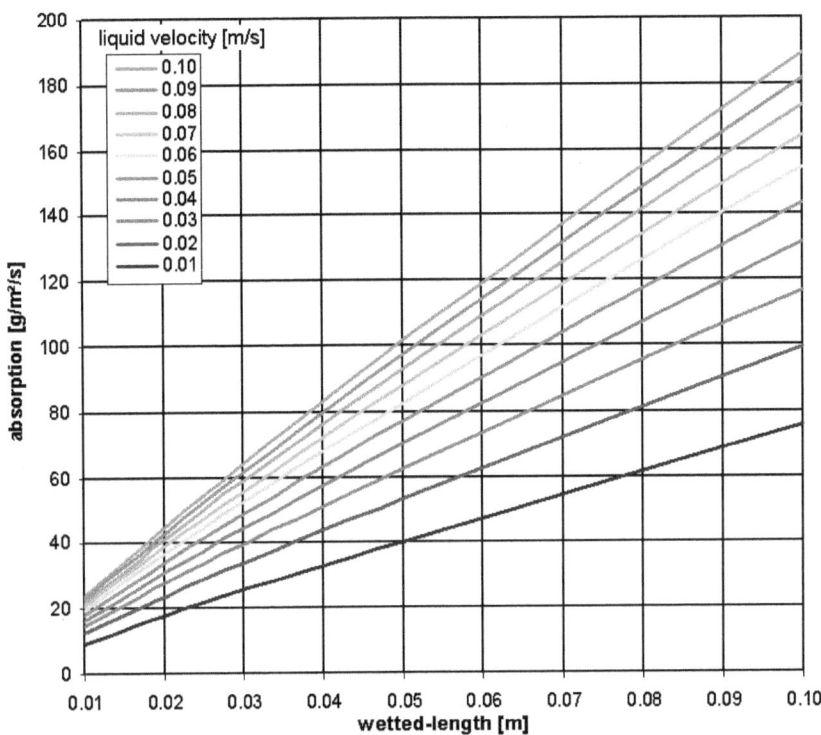

Figure 63. Example 14 Absorption Rate vs. Wetted Length

Comparing these last two figures reveals that the absorption rate is not linear with liquid velocity but is fairly linear with wetted length.

Chapter 10. Supersaturation

We next consider absorption of nitrogen in water beyond the saturation limit. As this is not a stable state, it cannot persist. It does occur and in one particular circumstance, has serious consequences. The case in point is a plunge pool where water cascades off a dam into the downstream. The constant churning of the water and air drives the concentration of nitrogen beyond the saturation limit. This unusual condition causes a sometimes-lethal reaction in fish similar to the *bends* experienced by divers ascending too rapidly. This problem was identified by the Army Corps of Engineers (USACE).

Development and utilization of a numerical model of the nitrogen supersaturation process became an important component of the Jennings Randolph project, serving to guide modifications and protect the fish. The earliest model was developed by the U. S. Army Corps of Engineers (USACE) Waterways Experiment Station (WES).[34]

The WES model is based on elementary principles of fluid flow and mass transfer with empirical correlations providing closure. The model is zero-dimensional, that is, it does not subdivide the domain into elements and solve conservation equations within the elements. Application of the model, therefore, is limited to configurations similar to those used to develop the empirical correlations, which provide model closure. The basic concept is that of a swarm of bubbles created by the plunging water rising to the surface and transferring nitrogen into the water in the stilling basin. Supersaturation can occur; because the bubbles are plunged downward into the water and experience greater than atmospheric pressure due to the hydrostatic pressure of the receiving water.

Geldert et al. began with a predictive model for dissolved gas levels downstream of a spillway developed by Roesner and Norton.[35] This model was based on a simple mass transfer model that can be expressed as Equation 10.1:

$$C_d = C_s - (C_s - C_u) e^{-Kt} \qquad (10.1)$$

---

[34] Geldert, D. A., Gulliver, J. S., and Wilhelms, S. C., "Modeling Dissolved Gas Supersaturation Below Spillway Plunge Pools," *Journal of Hydraulic Engineering*, May, 1998.

[35] Roesner, L. A., and W. R. Norton, "A Nitrogen Gas (N2) Model for the Lower Columbia River," Report No. 1-350, Water Resources Engineers, Inc., Walnut Creek, California, 1971.

where C_d is the downstream concentration, C_s is the saturation concentration, C_u is the upstream concentration, K is the mass transfer coefficient, and t is the residence time in the stilling basin. This equation formed the basis of the WES model.

Hibbs and Gulliver[36] utilized this next equation to calculate the effective saturation concentration, C_e:

$$C_e = C_s \left(1 + \frac{d_e \gamma}{P_a} \right) \qquad (10.2)$$

where C_s is the saturation concentration (taken to be 100%), d_e is the effective bubble depth, γ is the specific weight of water, and P_a is the atmospheric pressure. Geldert et al. provide three field data sets: Ice Harbor, The Dalles, and Little Goose. The measured effective saturation concentrations and the values computed using Equation 10.2 are shown in the figure below.

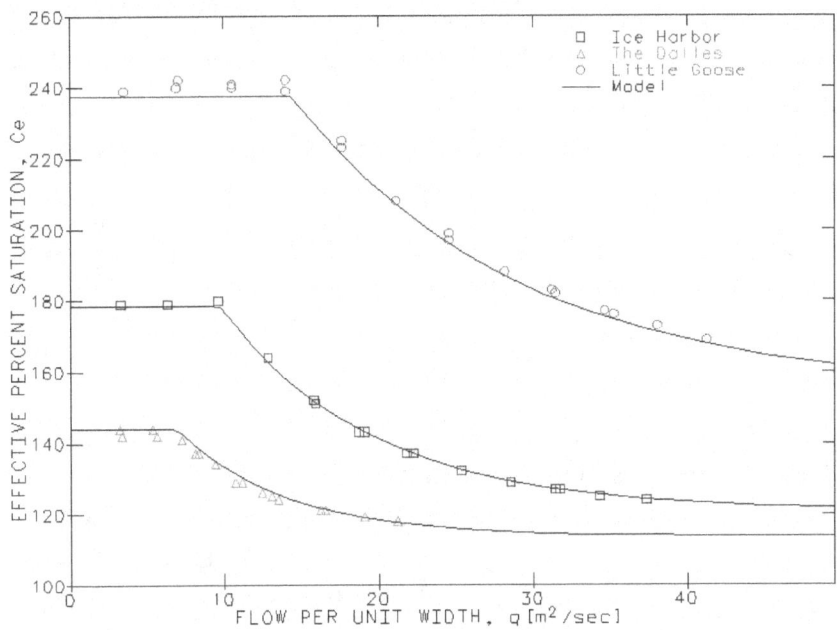

Figure 64. Computed and Measured Effective Saturation Concentration

[36] Hibbs, D. E. and J. S. Gulliver, "Prediction of an Effective Saturation Concentration at Spillway Plunge Pools." *Journal of Hydraulic Engineering*, Vol. 1, No. 3, pp. 940-949, 1977.

The effective depth is computed from the bubble half-life depth (i.e., the length traveled over the half-life), h_b, by Equation 10.3:

$$d_e = h_2 + (h_1 - h_2)e^{\left(1 - \frac{\beta h_b}{L_s}\right)} \quad for \quad \frac{\beta h_b}{L_s} > 1$$

$$d_e = h_1 \quad for \quad \frac{\beta h_b}{L_s} \leq 1 \tag{10.3}$$

where β is an empirical constant equal to 2.2, h_1 is the effective bubble depth in the stilling basin (presumed to be 2/3 of the stilling basin depth, h_s), h_2 is the effective bubble depth in the river (presumed to be 1/2 of the river depth, h_r), and L_s is the length of the stilling basin. The bubble half-life depth is computed from the discharge per unit width, q, and the bubble rise velocity, v_r (presumed be constant at 0.25 meters/second), by Equation 10.4.

$$h_b = \frac{q}{v_r}\ln(2) \tag{10.4}$$

Geldert et al. reasoned that the mass transfer included a bubble component into the water and a surface component out of the water. The rate of change of the concentration, C, is then given by Equation 10.5:

$$\frac{dC}{dt} = K_L a_b (C_e - C) + K_L a_s (C_s - C) \tag{10.5}$$

where K_L is the mass transfer coefficient, a_b is the bubble interfacial area per unit volume and a_s is the surface interfacial area per unit volume. The solution of this differential equation is given by Equation 10.6.

$$C_d = C_e - (C_e - C_u)\Lambda\Omega$$

$$\Lambda = e^{-(K_L a_b t_b + K_L a_s t_s)} + \frac{K_L a_s t_s}{K_L a_b t_b + K_L a_s t_s}\left(\frac{C_e - C_s}{C_e - C_u}\right) \tag{10.6}$$

$$\Omega = 1 - e^{-(K_L a_b t_b + K_L a_s t_s)}$$

where t_b is the residence time for the bubbles and t_s is the exposure time for the surface transfer. Geldert et al. presumed that the combination $K_L a_s t_s$ would be a dimensionless constant on the order of 1 for any particular application.

The void fraction, φ, is computed using Equation 10.7

$$\phi = \frac{v_j \lambda}{v_j \lambda + q} \tag{10.7}$$

where λ is an empirical constant on the order of 0.2 meters and v_j is the effective velocity of the plunging jet of water. Geldert et al. did not provide a means of obtaining v_j, simply stating that this was "computed by a standard water surface

80

profile technique." Geldert et al. used the void fraction and an empirical correlation to obtain the dimensionless bubble transfer group, $K_L a_b t_b$, given by Equation 10.8.

$$K_L a_b t_b = \alpha\phi \frac{(1-\phi)^{1/2}}{\left(1-\phi^{5/3}\right)^{1/4}} W_e^{3/5} R_q^{2/3} S_c^{-1/2} R_r^{-1} \tag{10.8}$$

where α is an empirical constant on the order of 1, W_e is the Weber number (Equation 10.9), R_q is the Reynolds number for the flow (Equation 10.10), S_c is the Schmidt number for air/water (Equation 10.11), and R_r is the Reynolds number for the rising bubbles (Equation 10.12).

$$W_e = \frac{\rho q^2}{\sigma d_j} \tag{10.9}$$

where ρ is the density of water, σ is the surface tension, and d_j is the effective depth of the plunging jet ($d_j=q/v_j$).

$$R_q = \frac{q}{\upsilon} \tag{10.10}$$

where υ is the kinematic viscosity of water.

$$S_c = \frac{V}{D} \tag{10.11}$$

where D is the air/water diffusion coefficient.

$$R_r = \frac{2 d_e v_r}{\upsilon} \tag{10.12}$$

These equations form the original WES model. Modifications are required in order to complete the model and to obtain good agreement with field data. The original WES model lacks an explicit calculation for the plunging jet velocity, v_j. In order to fill this gap in the model, a computer program was developed to "back out" the jet velocity implied by the data points for the three sites given in the WES report. These values were then compared to all of the dimensionless quantities, which can be formed from the site parameters. The best correlation obtained ($r^2=0.92$) is given in Equation 10.13 and illustrated in Figure 66.

$$\frac{v_j}{\sqrt{g\,h_t}} = 0.15 \left(\frac{\dfrac{q}{h_t}}{\sqrt{g\,h_t}} \right)^{0.23} \tag{10.13}$$

where g is the gravitational acceleration and h_t is the total (effective) head.

81

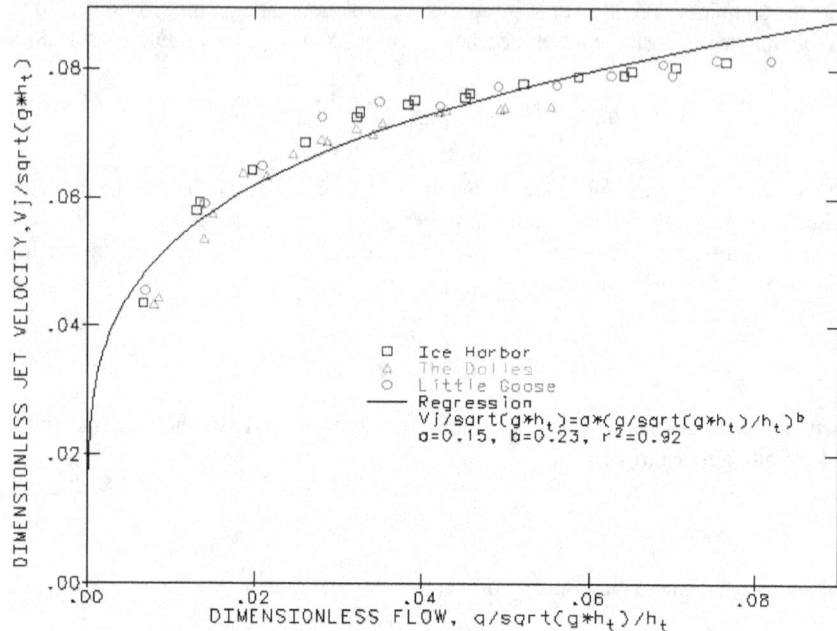

Figure 65. Dimensionless Correlation for Jet Velocity

The bubble transfer group, $K_La_bt_b$, can then be computed from Equation 10.8 and this correlation for the jet velocity (Equation 10.13). The results are illustrated in Figure 66. This same model can be applied to the Jennings Randolph site. The data and model results for all 4 sites are illustrated in Figure 67. Results for all four sites are shown in Figure 68.

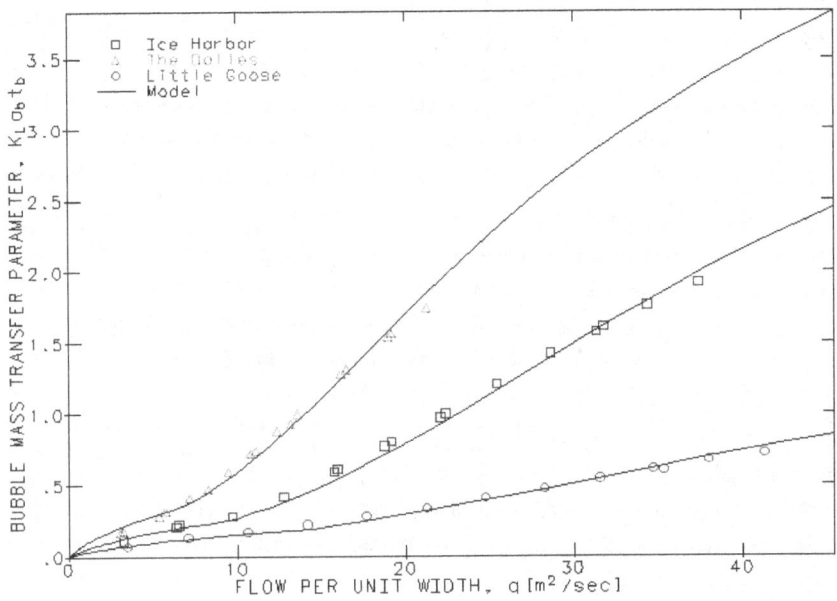

Figure 66. Bubble Transfer Group

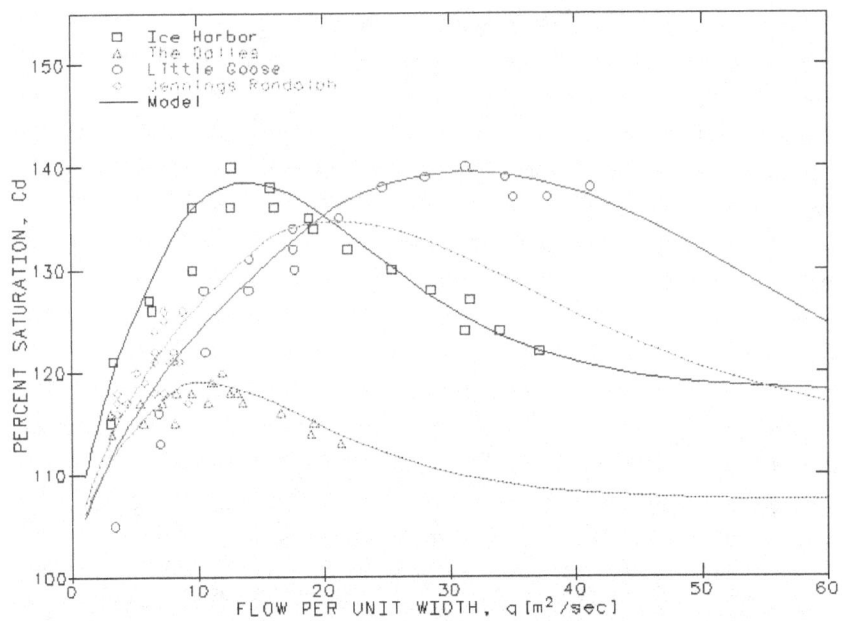

Figure 67. Data and Model Results for 4 Sites

As stated previously, this is a zero-dimensional empirical model. The Modified WES Model is significantly different than a one-, two-, or three-dimensional finite difference or finite element model in which the domain is subdivided into computational cells. Any direct applications of this model are limited to the variables which appear in the various equations, for instance, a single value must represent the depth of the stilling basin, h_s. If the depth of the stilling basin changes significantly over its length, this model will only accommodate a single number for the average or effective depth. If diverters or partitions are present in the stilling basin, there is no direct way to account for these in this model. There is also no direct way to account for such things as turbulence enhancers.

The equations could be modified to account for some changes in the basic configuration on which the model is based; but such modifications would require some theoretical basis and experimental data or some established scaling law. Large changes to the configuration, such as weirs, would require at least an additional model and are likely inherently incompatible with the Modified WES Model. The decision to use a zero-dimensional empirical model for this project was made during contract negotiations; as a multi-dimensional model would have required a significantly greater budget. Any multi-dimensional modeling would be outside the current scope of work.

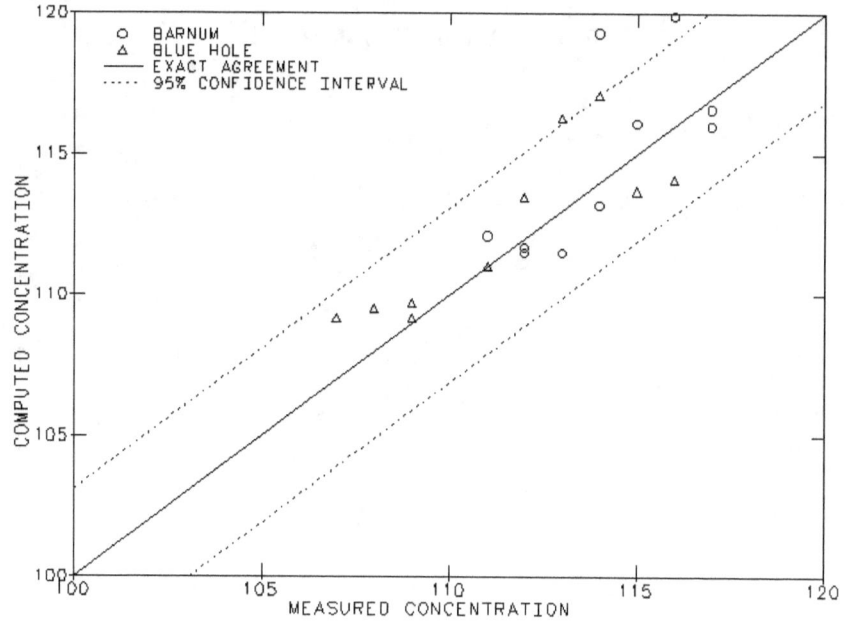

Figure 68. Agreement of Downstream Module and Field Data

The only way to incorporate the impact of structures such as weirs is to subdivide the domain into modules. The model must already be subdivided into the stilling basin module and the downstream module. The same equation for surface mass transfer is used for the stilling basin and the downstream modules; so this subdivision does not represent a second model. An aeration weir is modeled by inserting a weir module between the stilling basin and the downstream module. The coupling of these modules is limited to their impact on the inlet conditions to the next model, that is, the exit conditions from the stilling basin module become the inlet conditions to the weir module; and the exit conditions from the weir model become the inlet conditions to the downstream module.

The concentration downstream of the stilling basin is modeled using the surface mass transfer component only from the stilling basin model. This can be expressed as Equation 10.14. The agreement of the Downstream Module with field data is shown in Figure 68. This figure shows computed vs. measured concentrations. Data taken at Barnum are indicated by circles and data taken at Blue Hole are indicated by triangles. Points falling on the diagonal solid line would indicate exact agreement between the model and data. Points falling below the diagonal solid line indicate a model prediction less than the measured value. Points lying above the diagonal solid line indicate a model prediction greater than the measured value. The 95% confidence interval is indicated by the two diagonal dotted lines. This interval is a statistical measure of the accuracy of the Downstream Module and is equal to ±3.1%. This means that 95 out of 100 data points should be within 3.1% of the corresponding calculated value.

$$C_d = C_u + (C_s - C_u)\left(1 - e^{-K_L a_s t_s}\right) \tag{10.14}$$

The aeration weir is also a zero-dimensional model. A single value of effectiveness is used. The effectiveness is expected to change with flow. The concentrations upstream and downstream of the aeration weir are related by Equation 10.15.

$$C_d = C_u + \varepsilon (C_s - C_u) \tag{10.15}$$

where ε is the effectiveness. The range of ε is zero to one, where zero would mean no effect and one would be complete approach to saturation.

The Model can now be used to predict the impact on downstream supersaturation of changes in the basic geometry of the stilling basin. Figure 69 shows the impact of increasing or decreasing the depth of the stilling basin by a factor of 2. Also shown in Figure 69 are the predicted concentrations for a discharge of 6300 cfs (i.e., the largest operating point in the field data set). The Model predicts a saturation at the downstream end of the stilling basin of 125.6%. The Model also predicts that this value would increase to 128.1% if the depth of the stilling basin were doubled and decrease to 121.5% if the depth of the stilling basin were half of its current value. Figure 69 also shows the maximum discharge such that the saturation is no more than 110%. The predicted value is approximately

1420 cfs, 1260 cfs, and 1210 cfs, respectively, for the 3 cases. While this change is not insignificant, it is insufficient to ameliorate the problem and still allow for a reasonable operating range.

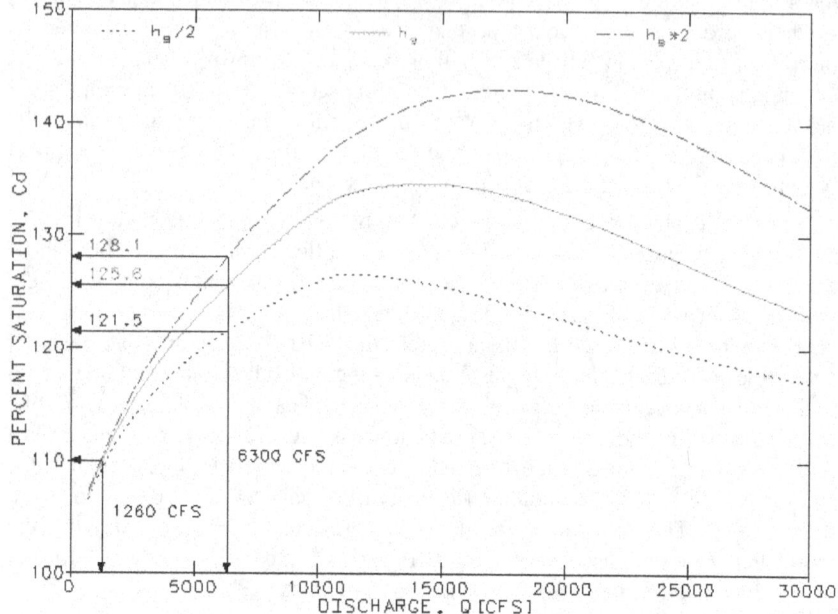

Figure 69. Impact of Stilling Basin Depth

Figure 70 shows the predicted impact of *river depth at the end of the stilling basin* on saturation. As seen in this figure, model predictions indicate that any modification of the river depth at this point would provide no benefit within the operating range.

86

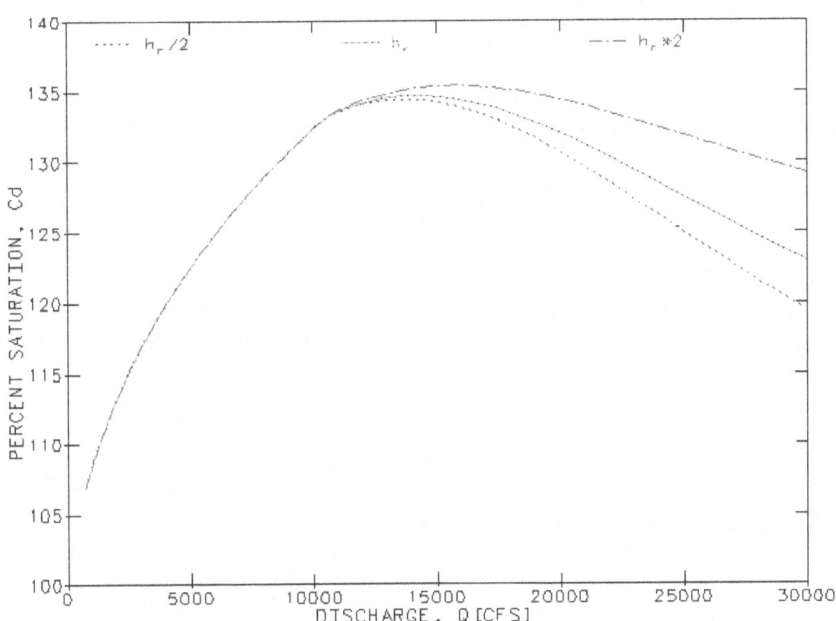

Figure 70. Impact of River Depth

Figure 71 shows the predicted impact of the *length of the stilling basin* on saturation. As was the case with a change in river depth, model predictions indicate that any modification of the stilling basin length would provide no benefit within the operating range.

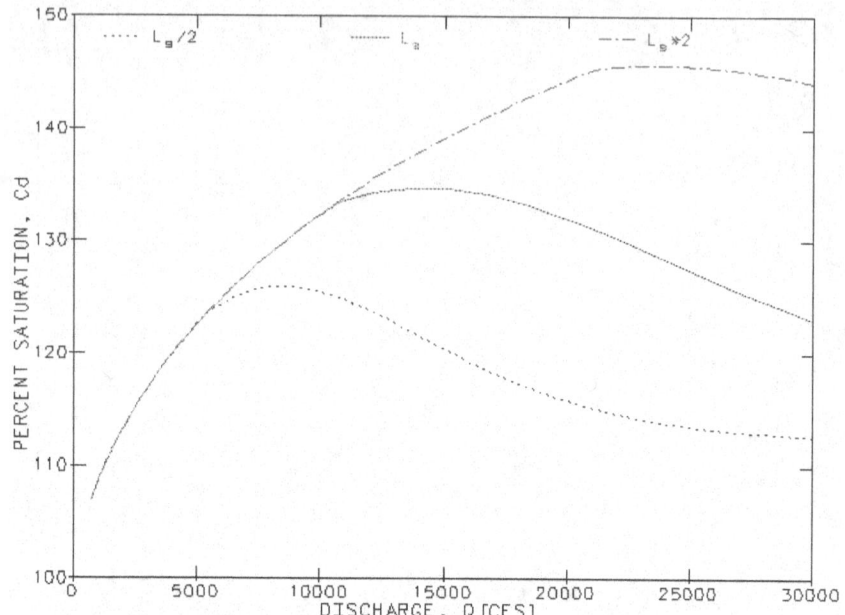

Figure 71. Impact of Stilling Basin Length

The Model does not accommodate a direct calculation for the *impact of a deflector*; however, this impact is estimated by limiting the effective bubble depth, d_e, which is an intermediate calculated parameter in the model, to some maximum plunge depth, d_p. The energy dissipated in the stilling basin would be the same whether or not a deflector were present. If the plunging water were deflected, one would expect greater turbulence in the stilling basin. In order to make some accommodation for this increased turbulence, the mass transfer coefficient, K_L, is increased by the same factor as the plunge depth is decreased. The results of these calculations are shown in Figure 72.

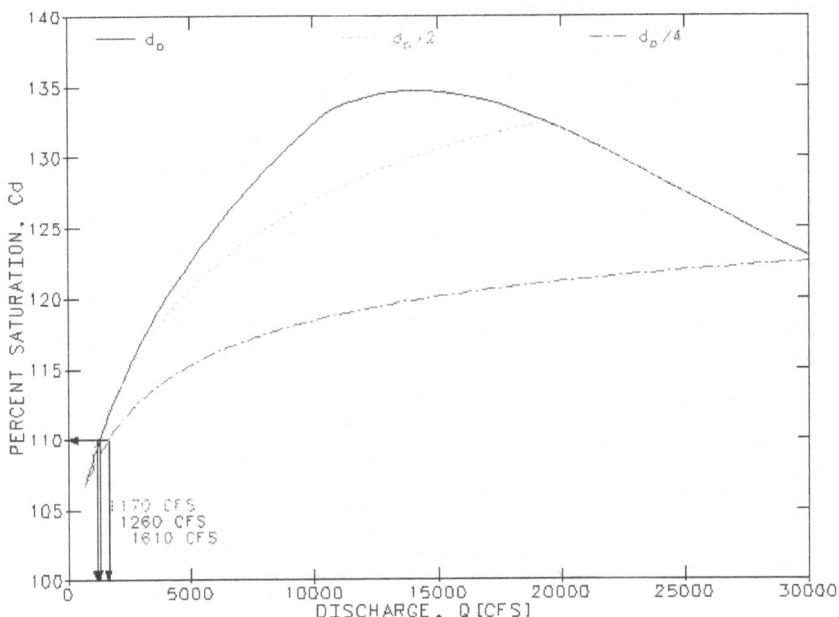

Figure 72. Impact of Limited Plunge Depth

These calculations must be considered an *estimate and cannot be ascribed the same level of confidence* as the basic model calculations. Figure 72 shows the model predicts that the greatest impact of limiting the plunge depth would be realized at discharges beyond the operating range. Figure 72 also shows that there may be a slight increase in supersaturation for low discharges were the plunge depth limited to one-half of its current value and a decrease in supersaturation were the plunge depth limited to one-fourth of its current value. This can be seen in Figure 72 by noting that the green dotted line (labeled $d_p/2$) is slightly above the solid blue line (labeled d_p); and the red chained line (labeled $d_p/4$) is noticeably below the solid blue line where the lines intersect $C_d=110\%$.

This seemingly anomalous impact at low discharge should not be considered significant; as this difference is less than the reasonable accuracy of the Model. The impact of limiting the plunge depth is clearly evident for larger discharges. Also shown on Figure 72 are the discharges corresponding to a saturation of 110%. The Model indicates that limiting the plunge depth to one-fourth of its current value would only increase the operating range from 1260 cfs to 1610 cfs, so as to not exceed 110% saturation. While this change is not insignificant, it is also insufficient to ameliorate the problem and still allow for a reasonable operating range.

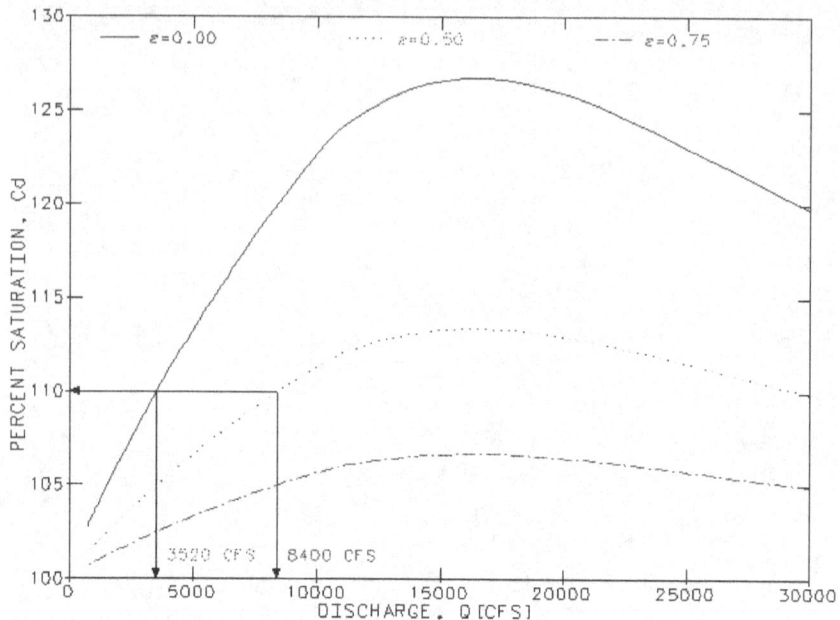

Figure 73. Computed Concentration 3 Miles Downstream

A Composite Model is used to estimate the impact of an aeration weir on downstream saturations. The basic Model (Equations 10.2 through 10.13) plus the Weir Module (Equation 10.15) plus the Downstream Module (Equation 10.14) can be used to estimate the saturation at any distance downstream. Figure 73 shows the computed concentrations at a distance 3 miles downstream. The aeration weir effectiveness, ε, is varied from 0 (or no weir), to 0.5 (a 50% effective weir), and 0.75 (a 75% effective weir). Of course, a completely effective weir ($\varepsilon=1$) would eliminate all downstream supersaturation.

Figure 73 shows that no weir is required ($\varepsilon=0$) in order to keep saturation levels at or below 110% for discharges up to 3520 cfs. A 50% effective weir ($\varepsilon=0.5$) would allow operation up to 8400 cfs without exceeding 110% at a distance 3 miles downstream. A 75% effective weir ($\varepsilon=0.75$) would allow operation at any discharge without exceeding 110% at a distance 3 miles downstream. The Model indicates that operation above 2400 cfs will result in saturations exceeding 110% at Barnum unless an aeration weir is installed. This saturation would be reached at Blue Hole for discharges exceeding 3180 cfs. If a 50% effective weir were installed, the discharge could reach 3620 cfs before the saturation would exceed 110% at Barnum and could reach 7800 cfs before the saturation would exceed 110% at Blue Hole. A 75% effective weir would allow any discharge without exceeding a saturation of 110% at Barnum or Blue Hole.

A modified version of the WES model for dissolved gas supersaturation below spillway plunge pools has been used to predict the impact of various changes to the stilling basin and adjacent riverbed on supersaturation. The Model was first calibrated using field data and then utilized to estimate the impacts. The Model predicts that only a slight benefit would be gained by reducing the depth of the stilling basin to one-half of its current value. The Model predicts that no benefit would be gained within the operating range by reducing the depth of the river at the end of the stilling basin to one-half of its current value. The Model predicts that no benefit would be gained within the operating range by doubling the length of the stilling basin. The Model predicts that only a slight benefit would be gained within the operating range by installing a deflector to limit the plunge depth to one-fourth of its current value.

In summary, none of these four modifications would provide the desired benefit of reducing the saturation level at the end of the stilling basin to no more than 110% over any significant portion of the operating range. The Model predicts that a 50% effective aeration weir would allow discharges up to 2400 cfs without exceeding a saturation of 110% downstream at Barnum or Blue Hole. The Model predicts that a 75% effective aeration weir would allow any discharge without exceeding a saturation of 110% downstream at Barnum or Blue Hole. Nothing short of an 85% effective aeration weir would keep the saturation levels at or below 110% between the weir and Barnum.

Chapter 11. Sorption

We next consider the migration of one chemical species (polychlorinated biphenyls or PCBs) through another species (water) in a media (soil) that participates in the process such that the combination is more complicated than any two by themselves. This contaminant (PCBs) has been a major concern for decades. Cleanup efforts have been extremely expensive and not always successful. This problem was identified by the US Environmental Protection Agency (USEPA).

Modeling of the transport of contaminants in groundwater has many applications. Groundwater systems are often quite complex; and many sophisticated numerical models are available. A sophisticated numerical model is not always practical or necessary. A simplified approach may provide sufficient information for management strategies or provide a basis for the selection of a sophisticated numerical model. A simplified approach based on transient one-dimensional diffusion in a finite medium is expanded to incorporate statistical measures of grain size in the medium and simple advection.

The model described in this chapter was developed beginning with the most simple concept of diffusion in an infinite Cartesian coordinate system and then transformed into a semi-finite Cartesian coordinate system. The length of the finite sub-domain was found to be critical. The model was then transformed from Cartesian coordinates to spherical in order to model granular media. The distribution of grain size was also included. The transport is separated into two distinct processes: diffusion and advection. The intra-grain sorption process is assumed to be diffusion-dominated; whereas, the inter-grain transport is assumed to be advection-dominated. In this model diffusion limits how much of the contaminant is available for advection. Contaminated sediment is modeled as a distributed source. The model is first developed without consideration of saturation or competition for receptor sites, which would be a worst-case scenario; as both of these effects would tend to diminish down-gradient concentration. Saturation and competition for receptor sites will be added subsequently.

The simplest model for contaminant transport would be one-dimensional transient diffusion in an infinite medium having uniform properties. The governing partial differential equation for mass transfer is given by Equation 1.3. For the purposes of this development, the medium will be divided at $x=0$. At time, $t=0$, the medium for $x<0$ is initially contaminated with a uniform concentration of $C=C_0$, and for $x>0$ with a uniform concentration of $C=0$. The medium extends infinitely far along x. The solution to this problem is given by Equation 4.1 and shown in the following figure:

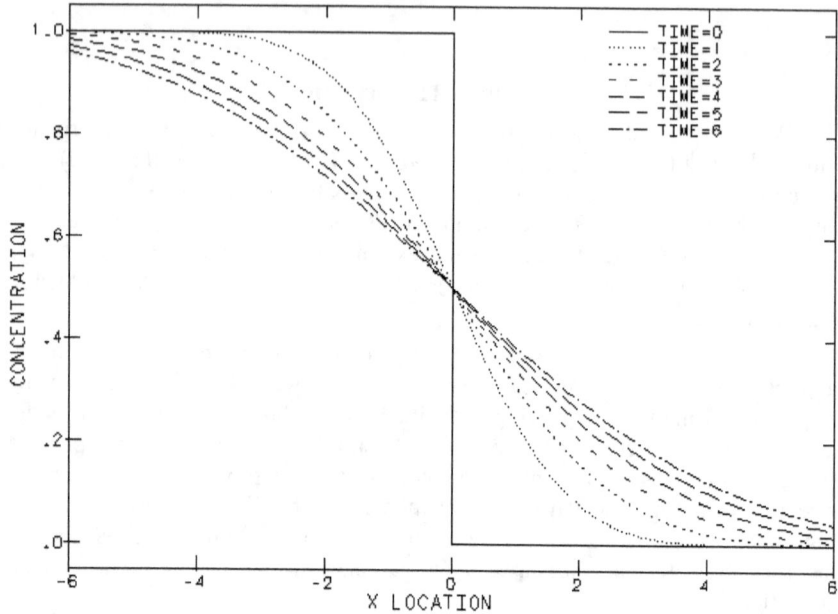

Figure 74. Concentration Profiles for Infinite Medium

It is clear from Equation 4.1 and Figure 74 that the ultimate profile at infinite time is a uniform concentration of 0.5. The initially contaminated medium for all $x<0$ constitutes an infinite source; whereas, the initially uncontaminated medium for all $x>0$ constitutes an infinite sink. Practical applications must consider a finite source. As the one-dimensional transient diffusion equation for an infinite medium having uniform properties is linear, the solution for a finite medium can be constructed using superposition. The actual medium is approximately semi-finite, that is, the initially contaminated zone is finite and the initially uncontaminated zone is infinite. In this case the applicable superposition is a reflection at $x=\pm R$. This makes the initial concentration, $C=0$ for $x<-R$, $C=C_0$ for $-R<x<R$, and $C=0$ for $x>R$. The superimposed solution is given by Equation 4.3. Concentration profiles at various times predicted by Equation 4.3 are shown in Figure 12, where $C_0=1$ and $R=3$.

Figure 12 shows the peak concentration (which is always at $x=0$) diminishing with time and the concentration profile flattening out as the finite initial contamination diffuses out into the infinite medium. Another result of this modification is illustrated in Figure 75, which shows the concentration at a single point, $x=2$, over time. The concentration predicted for an infinite source

94

(Equation 1.3) is shown by the dotted curve. The concentration predicted for a finite source (Equation 4.1) is shown by the solid curve. The infinite source solution does not exhibit a maximum; instead, it asymptotically approaches *C=0.5*. The finite source solution exhibits a clear maximum. The timing and magnitude of the maximum is of particular interest when considering contaminant transport. Both are dependent on the distance to the reflection, *R*, or the effective length scale of the contaminated medium.

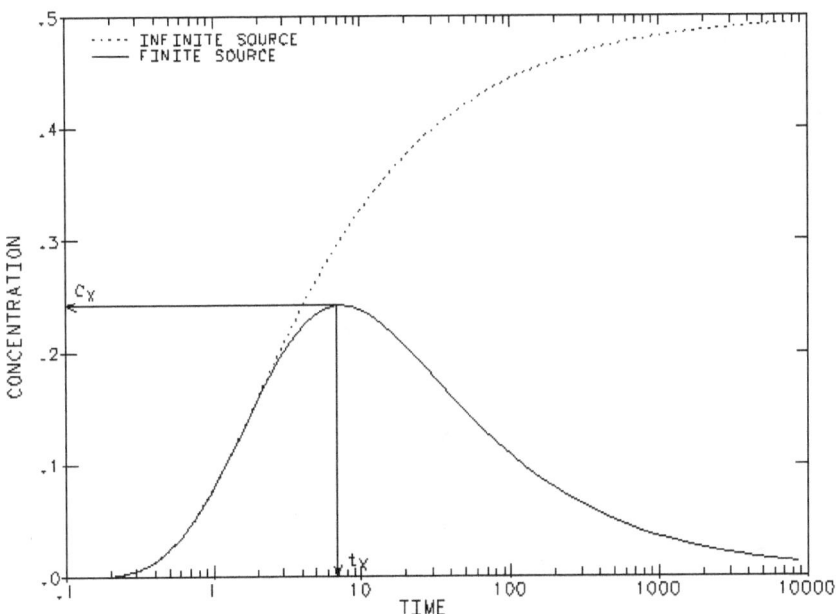

Figure 75. Concentration at a Point over Time

The time corresponding to the maximum concentration can be found by differentiating Equation 4.1 with respect to time and solving for the root, as in Equation 11.1. The root of Equation 11.1 is given by Equation 11.2. The maximum concentration over time is then found by using this value in Equation 4.3.

$$x \, e^{\left(\frac{-x^2}{4Dt}\right)} = (x + R) \, e^{\left[\frac{-(x+R)^2}{4Dt}\right]} \qquad (11.1)$$

$$t_X = \frac{R(2x + R)}{4 \, D \ln\left(\frac{x + R}{x}\right)} \qquad (11.2)$$

The maximum concentration and corresponding time are shown in Figures 76 and 77, respectively, for a range of *R*s. These figures show that the reflection

95

distance, **R**, or the effective length scale of the contaminated medium has considerable impact on the maximum concentration and corresponding time. This is why accurate characterization of the medium is critical in developing contamination management strategies.

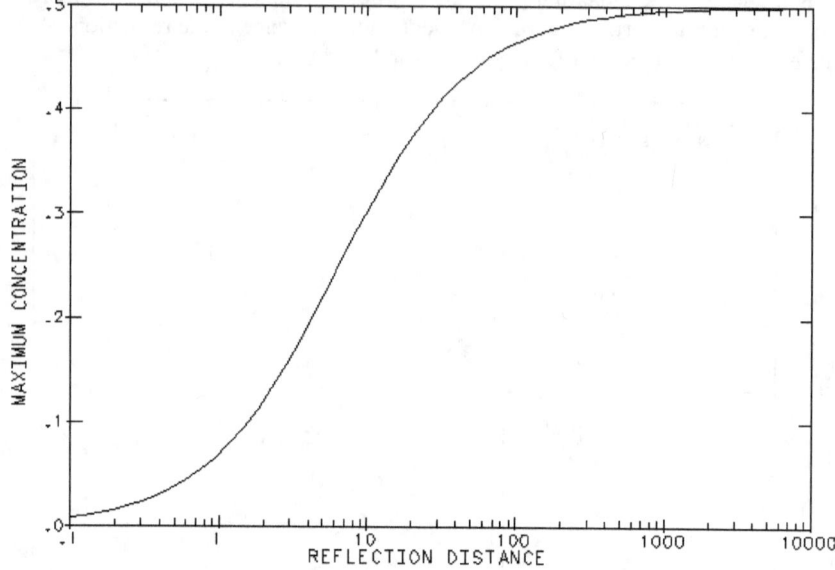

Figure 76. Maximum Concentration vs. Reflection Distance

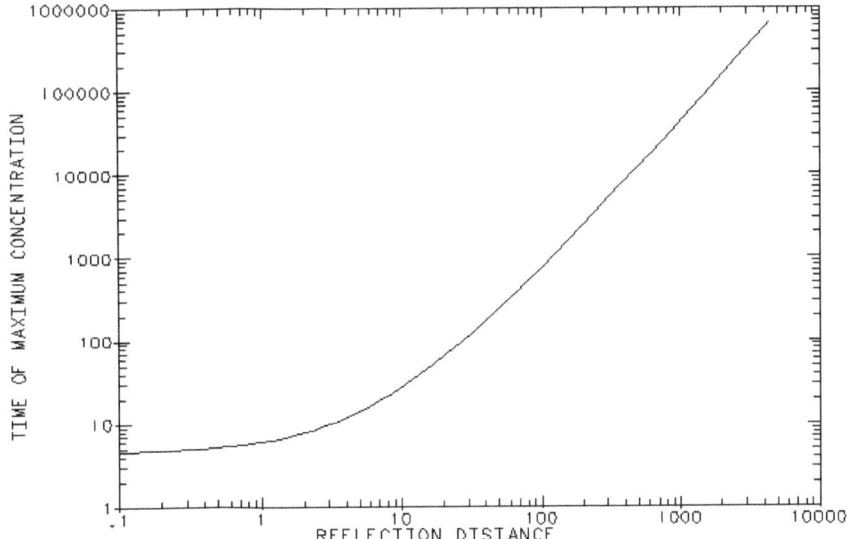

Figure 77. Time of Maximum Concentration vs. Reflection Distance

Treating the initially contaminated medium as a monolithic slab results in the largest maximum concentration and the longest time to reach that maximum concentration. A more realistic model for the initially contaminated medium is one composed of variably-sized spheres whose distribution of diameters can be described statistically. The governing partial differential equation for transient diffusion in spherical coordinates is given by Equation 11.3.

$$\frac{\partial C}{\partial t} = \frac{1}{r^2} \frac{\partial}{\partial r} \left[r^2 D \frac{\partial C}{\partial r} \right] \qquad (11.3)$$

Here r is the radial distance from the center of the grain. Equation 11.3, which leads to the solution, Equation 4.3. Figure 78 shows the same calculations as Figure 12 with spherical spreading and the current temporal variables.

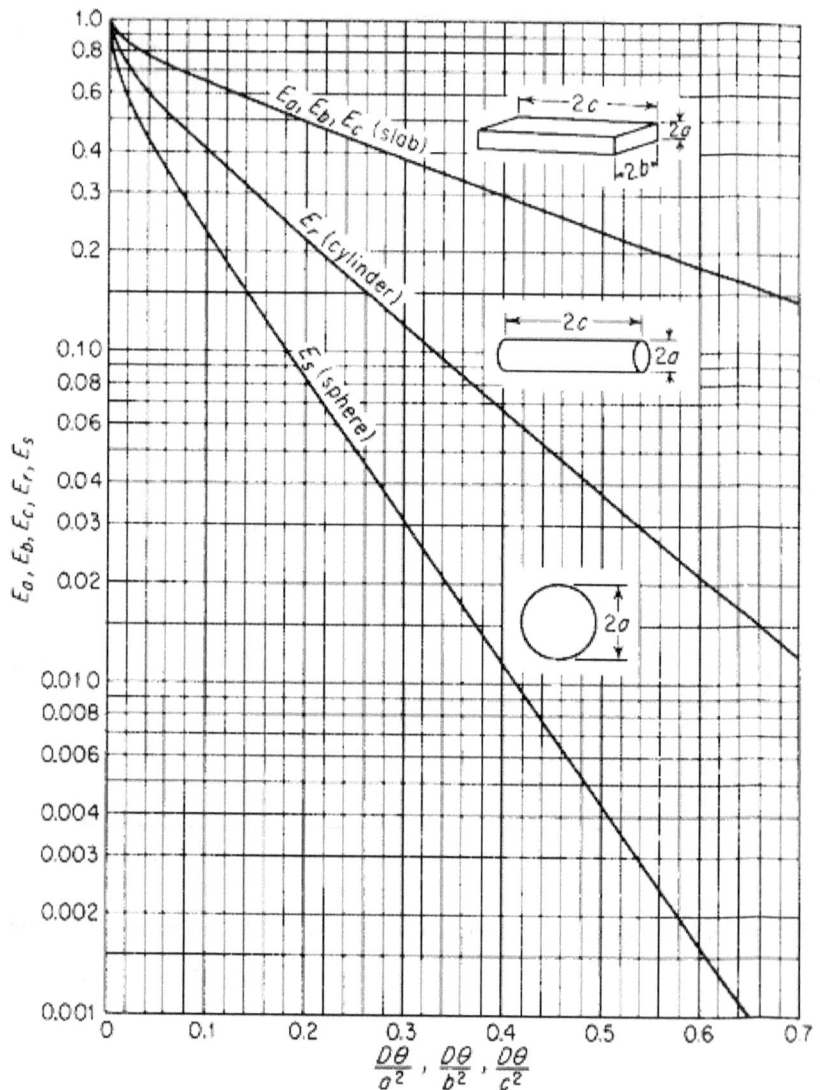

Figure 80. Unsteady-State Diffusion after Treybal

Figures 76 and 77 show that the maximum concentration and the time to reach that maximum depends on the grain radius. The model for the sorption process accounts for the variation with grain size through Equation 4.3. It is assumed that the distribution of grain sizes can be approximated through statistical means. Figure 81 shows grain diameter vs. probability for eight samples taken from the same area. Significant variability is evident even between samples taken in close proximity. This variability contributes to the uncertainty of the results.

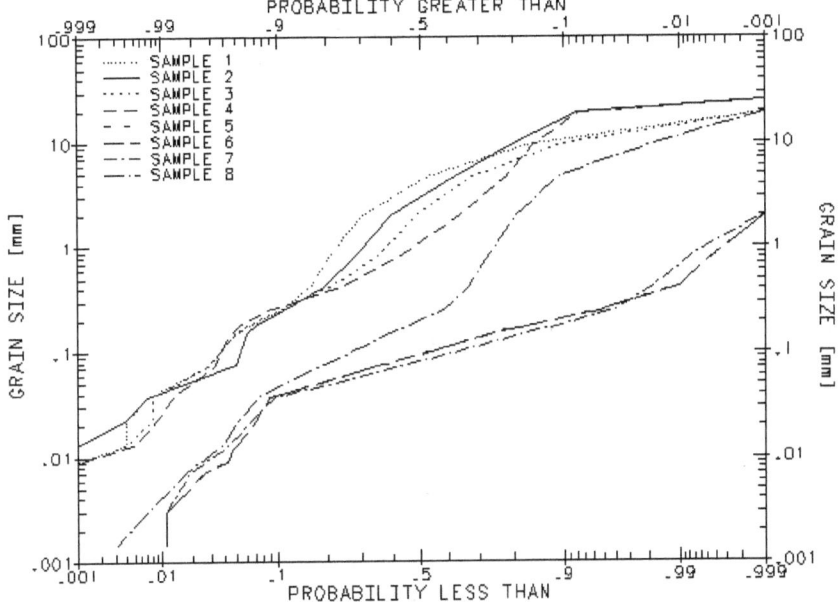

Figure 81. Measured Grain Size Probability for 8 Samples

A normal probability distribution produces a straight line on a probability axis scale as constructed in Figure 81. The value where the curve intersects a probability of 0.5 is the mean. The slope of the line is always positive and proportional to the standard deviation. The normal probability distribution function is given by Equation 11.4:

$$F(G) = \frac{e^{-\frac{1}{2}\left(\frac{G-G_M}{\sigma}\right)^2}}{\sigma\sqrt{2\pi}} \qquad (11.4)$$

Where G_M is the mean grain size, σ is the standard deviation, and $F(G)$ is the probability distribution of grain size G. The cumulative probability, or the axis in Figure 81 labeled *Probability Less Than*, for some grain size, G, is the integral of the probability distribution from $-\infty$ to G, or Equation 11.5, which can be resolved to the error function.

$$P(G) = \int_{-\infty}^{G} \frac{e^{-\frac{1}{2}\left(\frac{G-G_m}{\sigma}\right)^2}}{\sigma\sqrt{2\pi}} \, dG = \frac{1}{2}\left[1 + erf\left(\frac{G-G_M}{\sigma\sqrt{2}}\right)\right] \qquad (11.5)$$

101

where **P(G)** is the cumulative probability of grain size **G**. A best fit normal probability distribution is obtained for Sample 8 with a mean grain size of 0.30 mm (note: $0.30=10^{-0.53}$) and a log standard deviation of 0.79, as shown in Figure 82.

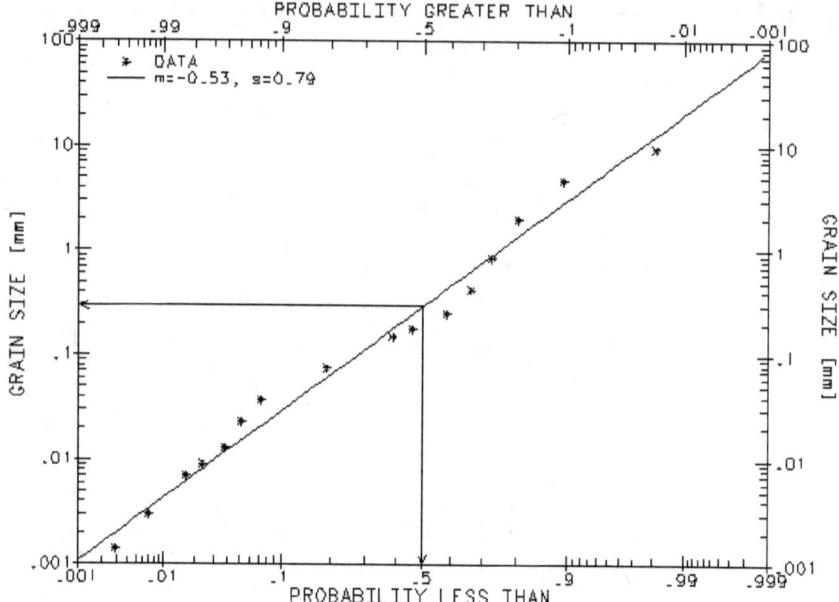

Figure 82. Best Fit Normal Probability Distribution for Sample 8

Equation 11.3 with uniform properties is linear; therefore, a statistical composite concentration involving many grains can be constructed by superposition using Equation 4.3. The result is given by Equation 11.6, which must be integrated numerically.

$$C(r,t)=\frac{C_0}{2}\int_{-infinity}^{infinity}\frac{e^{-\frac{1}{2}\left(\frac{G-G_M}{\sigma}\right)^2}}{\sigma\sqrt{2\pi}}\left[erfc\left(\frac{r-R}{2\sqrt{Dt}}\right)-erfc\left(\frac{r+R}{2\sqrt{Dt}}\right)\right]dG \quad (11.6)$$

Equation 11.6 forms the basis for analyzing practical problems of diffusion in granular porous media. The Sample 8 grain size data are used as an example (cf. Figure 81). The mean grain size is 0.30 mm, making the mean radius 0.015 cm. A typical diffusion coefficient of 0.00001 cm²/sec is used. Figure 83 is obtained using a grain size standard deviation of 0.079. The profiles in Figure 83 are quite similar to those in Figure 78, which is the slab results. This is an important check on the asymptotic behavior of Equation 11.6. If the grain size is relatively large and the standard deviation is relatively small (i.e., nearly uniform grains), compared to the diffusion coefficient and characteristic time, Equation 11.6 should approach the slab result, or Equation 2.1, which it does.

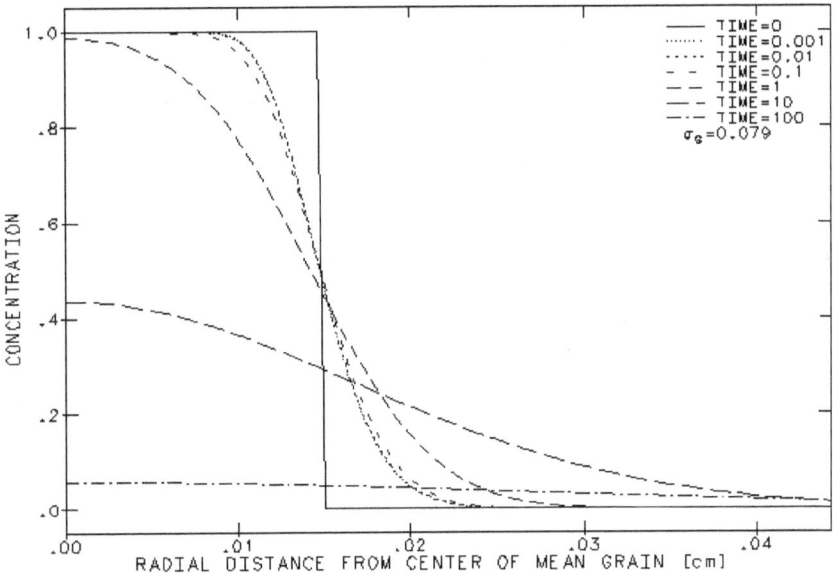

Figure 83. Statistically Averaged Spherical Conc. Profiles for σ=0.079

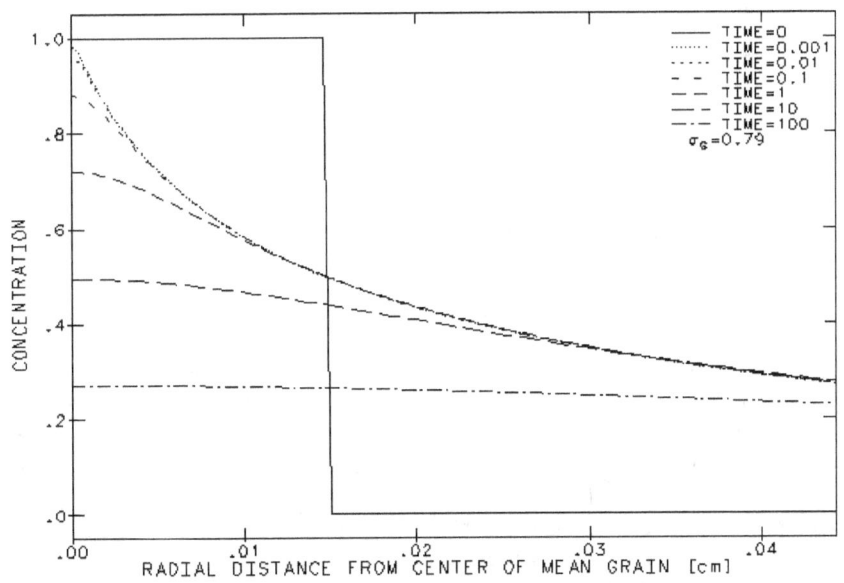

Figure 84. Statistically Averaged Spherical Conc. Profiles for σ=0.79

If the grain size standard deviation is increased to 0.79, a significant change in the shape of the profiles can be seen (cf. Figure 84). Increasing the grain size standard deviation increases the fraction of smaller and larger diameter grains relative to the mean. The contribution of the smaller grains to the integrated concentration is a more rapid decrease at the center; as the relative distance from the source is much larger. The contribution of the larger grains is to hold the concentration up farther away from the center; as the part of the concentration profile plotted is still deep within the large grains. Mass is conserved, thus, the area under each curve when weighted by the radius squared is the same. As more smaller grains *pull* the profiles down toward the center and more larger grains *stretch* the profiles out, a different shape is obtained (cf. Figure 84) as the radial effects dominate.

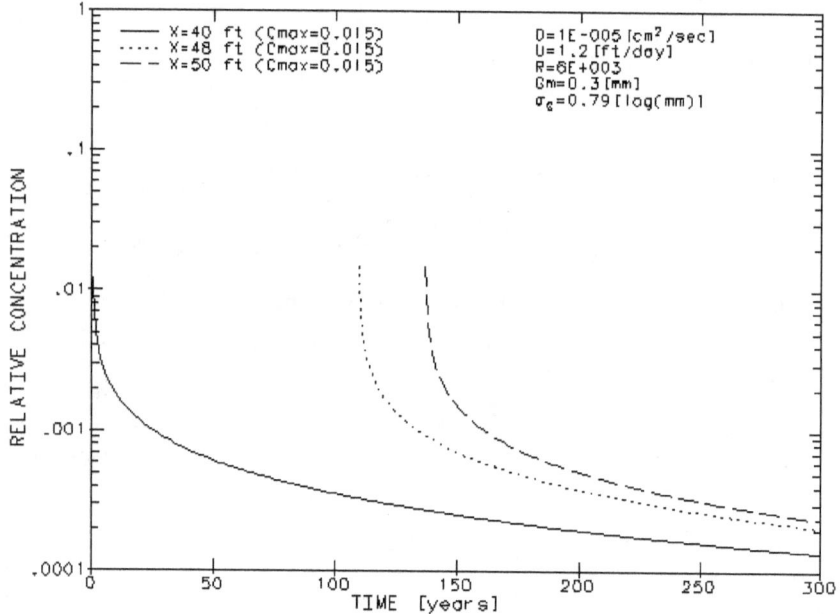

Figure 85. Application Results

These developments were applied to the following example. An originally contaminated zone of length, $L=1220$ cm (40 ft.) is adjacent to an originally uncontaminated zone of length 305 cm (10 ft.). Groundwater flows through the originally contaminated zone and into the originally uncontaminated zone. The advective velocity is 0.00042 cm/sec (1.2 ft./day). The effective retardation factor, $\mathcal{R}$, for the contaminant is 6000. The grain size distribution for Sample 8 was used. Superposition was used to arrive at Equation 11.17, which must be numerically integrated over the originally contaminated length.

104

$$C(X,t) = \int_0^L \left[C\left(R_M, \frac{t}{\Re} - \frac{X-L-x}{U} \right) - C\left(R_M, t - \frac{X-L-x}{U/\Re} \right) \right] \frac{d\,x}{L} \quad (11.17)$$

Where $C(r,t)$ is given by Equation 11.17 and X is the distance from the beginning of the initially contaminated zone to the point of interest, in this case, the end of the originally uncontaminated zone, or 1524 cm (50 ft.). Note that Equation 11.17 contains an indefinite integral (i.e., from $-\infty$ to $+\infty$) over the grain size; thus, Equation 11.18 contains a nested double integral which must be computed numerically. The first term in Equation 11.18 would apply if both zones were initially contaminated. The second term in Equation 11.18 accounts for the initially uncontaminated zone by removing the residual contribution of the first term in the uncontaminated zone.

The results of Equation 11.18 are shown in Figure 85. The contamination appears immediately at the interface between the two zones (i.e., $X=40$ ft); but most of it is taken into the uncontaminated grains; as the time scale for the intra-grain adsorption is much smaller than the time scale for the retarded advection. The abrupt appearance of contaminant at $X=48$ ft and $X=50$ ft is due to the assumption that the intra-grain transfer is diffusion-dominated and the inter-grain transport is advection-dominated. This is a type of worst-case scenario. In a physical system there would be some diffusion leading the advection, which would tend to round-off the leading edge of the curve, resulting in a smoothed peak, rather than an abrupt appearance. A summary of results for various input parameters is listed in Table 11.1. The arrival time is in years.

Table 11.1. Summary of Application Results

		bank soil			river sediment		
retardation factor	initial concen	arrival time	contam time	max concen	arrival time	contam time	max concen
6000	15	110	-NA-	0.22	137	-NA-	0.22
6000	50	110	-NA-	0.74	137	-NA-	0.74
6000	100	110	-NA-	1.49	137	-NA-	1.49
60000	15	1095	-NA-	0.22	1369	-NA-	0.22
60000	50	1095	-NA-	0.75	1369	-NA-	0.75
60000	100	1095	-NA-	1.49	1369	1369	1.49

Appendix A. Diffusion Coefficients

There are numerous sources for various properties available on the Web. Diffusion coefficients are published by a many organizations as well as government agencies, especially those that may pose a health hazard. The following tabulated values are typical.

Table A1. Pairs of Gases at 1 atm.

A	B	cm²/s	°C	A	B	cm²/s	°C
Ar	CO_2	0.200	20	H_2O	air	0.219	0
Ar	N_2	0.196	0	H_2O	CO_2	0.202	35
Ar	O_2	0.200	20	H_2O	H_2	1.020	35
CH_4	Air	0.149	0	H_2O	He	0.900	35
CO	CH_4	0.137	0	H_2O	N_2	0.256	35
CO	CO_2	0.192	0	He	Ar	0.641	0
CO	N_2	0.139	0	CCl_4	H_2	0.293	0
CO_2	air	0.178	0	CCl_4	O_2	0.064	0
CO_2	CH_4	0.144	0	CH_4	air	0.138	0
CO_2	N_2	0.116	0	$(C_2H_5)_2O$	air	0.079	0
CO_2	N_2O	0.153	0	$(C_2H_5)_2O$	CO_2	0.054	0
H_2	air	0.096	0	$(C_2H_5)_2O$	H_2	0.299	0
H_2	Ar	0.602	25	$(CH_3)_2CO$	H_2	0.361	0
H_2	C_2H_6	0.420	25	C_2H_5OH	air	0.099	0
H_2	CH_4	0.625	0	C_2H_5OH	CO_2	0.069	0
H_2	CO	0.601	0	C_2H_5OH	H_2	0.377	0
H_2	CO_2	0.550	0	C_6H_6	H_2	0.318	0
H_2	D_2	1.200	0	C_6H_6	O_2	0.080	0
H_2	N_2	0.674	0	$C_6H_4Cl_2$	air	0.069	35
H_2	N_2O	0.535	0	$C_{10}H_8$	air	0.059	35
H_2	O_2	0.697	0	Hg	N_2	0.119	0
H_2	SF_6	0.140	20	I	air	0.069	0
H_2	SO_2	0.480	0	I	N_2	0.070	0

For other temperatures, the diffusion coefficient for gases can be estimated as being approximately proportional to the absolute temperature raised to some power, n, between 1.75 and 2.00, as in the following:

$$D \propto T^n \tag{A.1}$$

All of the values in this appendix can be found in diffusion_coefficients.xls in the examples folder of the online archive accompanying this text.

Some diffusion coefficients are tabulated for a range of temperatures, as in the following table:

Table A2. D [cm²/s] in Air at 1atm.

gas	0°C	20°C	100°C	200°C	300°C	400°C
Ar	0.167	0.189	0.289	0.437	0.612	0.810
CH_4	0.149	0.210	0.321	0.485	0.678	0.899
CO	0.149	0.208	0.315	0.475	0.662	0.875
CO_2	0.110	0.160	0.252	0.390	0.549	0.728
H_2	0.668	0.756	1.153	1.747	2.444	3.238
H_2O	0.219	0.242	0.399	0.638	0.873	1.135
H_2S	0.028	0.032	0.050	0.079	0.114	0.154
He	0.617	0.697	1.057	1.594	2.221	2.933
SF_6	0.047	0.067	0.150	0.233	0.329	0.438
SO_2	0.036	0.041	0.065	0.102	0.147	0.200

Other approximations are also available, including Graham's[38] law (principle) of Effusion (or Diffusion) states that the rate of effusion of a gas is inversely proportional to the square root of the mass of its particles, which can be expressed:

$$\frac{R_1}{R_2} = \sqrt{\frac{M_2}{M_1}} \qquad (A.2)$$

where R_1 is the rate of effusion for the first gas, R_2 is the rate of effusion for the second gas, M_1 is the molar mass of the first gas, and M_2 is the molar mass of the second gas.

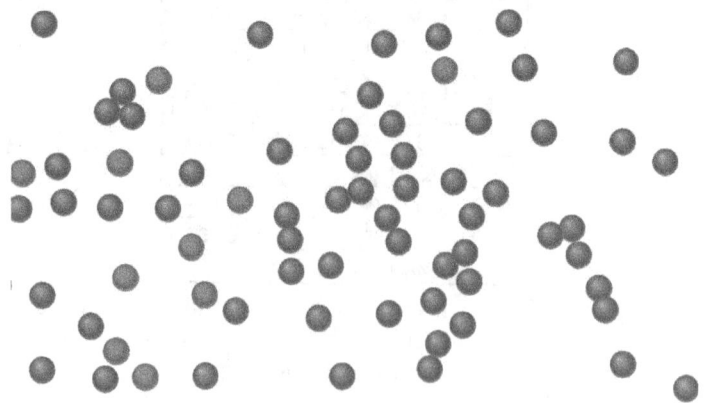

---

[38] Thomas Graham (1805–1869) Scottish chemist who studied dialysis and diffusion.

Diffusion coefficients for gases in liquids may be significantly different than for two gases, as illustrated in the following table:

Table A3. D [cm²/s] in Water (lqd.)

gas	10 °C	15 °C	20 °C	25 °C	30 °C	35 °C
Ar	2.28	2.36	2.43	2.50	2.57	2.65
C_2H_2	1.43	1.59	1.78	1.99	2.23	2.29
CH_3Br	1.23	1.27	1.31	1.35	1.39	1.43
CH_3Cl	1.28	1.32	1.36	1.40	1.44	1.48
CH_4	1.24	1.43	1.62	1.84	2.08	2.35
$CHCl_2F$	1.64	1.70	1.75	1.80	1.85	1.91
Cl_2	1.10	1.13	1.50	1.89	1.95	2.00
CO_2	1.26	1.45	1.67	1.91	2.17	2.47
H_2	3.62	4.08	4.58	5.11	5.69	6.31
H_2S	1.24	1.28	1.32	1.36	1.40	1.44
HBr	2.88	2.97	3.06	3.15	3.24	3.34
HCl	2.80	2.89	2.98	3.07	3.16	3.25
He	5.67	6.18	6.71	7.28	7.87	8.48
Kr	1.20	1.39	1.60	1.84	2.11	2.40
N_2	1.83	1.88	1.94	2.00	2.06	2.12
N_2O	1.37	1.62	2.11	2.57	3.00	3.48
Ne	2.93	3.27	3.64	4.03	4.45	4.89
NH_3	1.26	1.30	1.50	1.55	1.59	1.64
NO_2	0.80	0.94	1.23	1.40	1.59	1.85
O_2	1.41	1.67	2.01	2.42	2.82	3.28
Rn	0.81	0.96	1.13	1.33	1.55	1.80
SO_2	1.19	1.38	1.62	1.83	2.07	2.32
Xe	0.93	1.08	1.27	1.47	1.70	1.95

Diffusion coefficients for several substances in crystalline silicon can be approximated by the following formula:

$$\log_{10} D = s\left(\frac{10^4}{T}\right) + i \qquad (A.3)$$

where s is the slope and i is the intercept. The following table is based on graphical values provided by Mokhov.[39]

---

[39] Mokhov, E. N., "Doping of SiC Crystals during Sublimation Growth and Diffusion," in *Crystal Growth*, Glebovsky, Ed., IntechOpen, Rijeka, Russia, 2019.

Table A4. Diffusion in Silicon

name	abbrev.	slope	intercept
Aluminum (solid phase)	Alsp	-9.472	20.484
Aluminum (vapor phase)	Alv	-3.152	0.835
Beryllium (solid phase)	Bes	-2.537	0.642
Beryllium (vapor phase)	Bev	-1.871	0.232
Boron (solid phase)	Bsp	-7.690	16.073
Boron (vapor phase)	Bv	-2.961	1.829
Carbon	C	-4.432	8.847
Gallium	Ga	-2.931	-0.630
Lithium	Li	-1.713	1.008
Nitrogen	N	-6.665	11.408
Oxygen	O	-4.744	5.465
Phosphorous	P	-8.376	19.656
Silicon	Si	-4.890	6.920

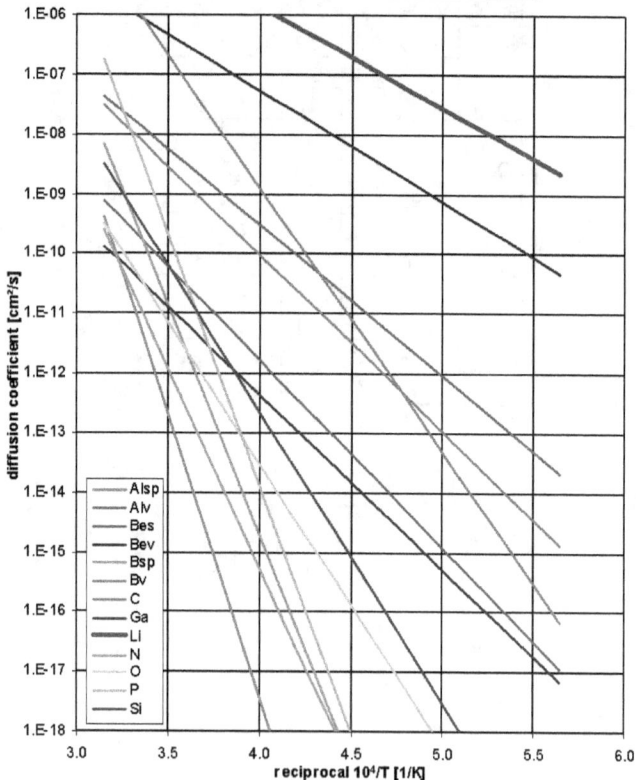

Figure 86. Diffusion Coefficients in Silicon

Appendix B: Gauss Error Function

The Gauss Error function is defined by Equation 4.2:

$$erf(x) = \frac{2}{\pi} \int_0^x e^{-t^2}\, dt \qquad (4.2)$$

and shown in the figure below:

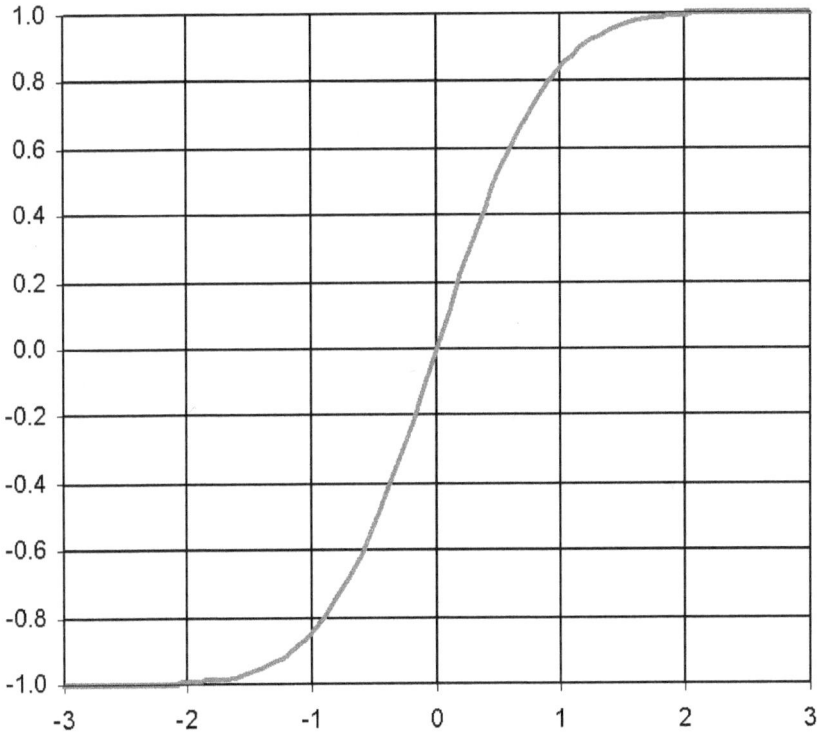

Figure 87. Gauss Error Function

The tails quickly approach −1 for x<−3 and +1 for x>+3. An approximation is provided by Abramowitz & Stegun in their *Handbook of Mathematical Functions*.[40] The Excel macro is listed below and may be found in example4.xls:

```
Option Explicit
Function erf(x As Double) As Double
    Dim e As Double, q As Double, t As Double, y As Double
    If (Abs(x) < 0.0000001) Then
      erf = 0
    ElseIf (x > 6) Then
      erf = 1
      Exit Function
    ElseIf (x < -6) Then
      erf = -1
      Exit Function
    Else
      y = Abs(x)
      t = 1 / (1 + 0.3275911 * y)
      q = (((((1.061405429 * t - 1.453152027) * t +
      1.421413741) * t - 0.284496736) * t + 0.254829592) *
      t
      e = 1 - q / Exp(x * x)
      If (x < 0) Then
        erf = -e
      Else
        erf = e
      End If
    End If
End Function
Function erfc(x As Double) As Double
    erfc = 1 - erf(x)
End Function
```

[40] Abramowitz, M. and I. A. Stegun, *Handbook of Mathematical Functions* first published by the National Bureau of Standards as Technical Monograph No. 55. This very useful reference may be obtained free on-line as a PDF from several different web sites.

Appendix C. Merkel's Equation

Merkel derived the following equation for demand:

$$\frac{KaV}{L} = C_{PW} \int_{T_{OUT}}^{T_{IN}} \frac{dT}{h_F - h_A} \qquad (8.15)$$

Merkel chose the 4-point Chebyshev method to integrate it, which isone of the simplest methods available. It can be expressed:

$$\int_a^b f(x)\,dx = \sum_{i=1}^{n} w_i f\big(a + x_i(b - a)\big) \qquad (C.1)$$

where w_i are the weights (or multipliers) and x_i are the abscissas (or points of evaluation). In this case $n=1$, $w_i=0.25$, and $x_i=0.1$, 0.4, 0.6, and 0.9. In all such forms of numerical integration the sum of the weights should be unity, $\Sigma w_i=1$. In most cases, the weights are not equal. Equal weights are unusual. The Chebbyshev method is notable in this respect. The following Excel macro calculates the integral above:

```
Function fMerkel(Twb As Double, Ran As Double,
    apr As Double, rLG As Double) As Double
Dim i As Integer, Ha As Double, Hain As Double,
    Haex As Double, Hw As Double
Dim Tcold As Double, Thot As Double, Tw As Double,
    X(4) As Double
X(1) = 0.1
X(2) = 0.4
X(3) = 0.6
X(4) = 0.9
Hain = fHtwb(Twb)
Haex = Hain + Ran * rLG
Tcold = Twb + apr
Thot = Tcold + Ran
fMerkel = 0#
For i = 1 To 4
  Tw = Tcold + X(i) * Ran
  Hw = fHtwb(Tw)
  Ha = Hain + X(i) * (Haex - Hain)
  If (Hw <= Ha) Then
    fMerkel = 999#
    Exit Function
  End If
  fMerkel = fMerkel + 0.25 / (Hw - Ha)
```

```
    Next i
    fMerkel = fMerkel * Ran
End Function
```

Appendix D. Cooling Tower Demand Curves

There are several options for calculating cooling tower demand curves not covered in this text. These include: 1) Merkel's method, 2) a more accurate method using fewer simplifying assumptions, and 3) orientation of the streams (counterflow or crossflow). You will find a program (DEMAND) on the web site listed in the forward that solves the demand and supply equations, comes with a Windows GUI, and creates graphics that can be copied and pasted into MS Word documents and Excel spreadsheets. Figure 41 is typical of counterflow and the figure below is typical of crossflow:

115

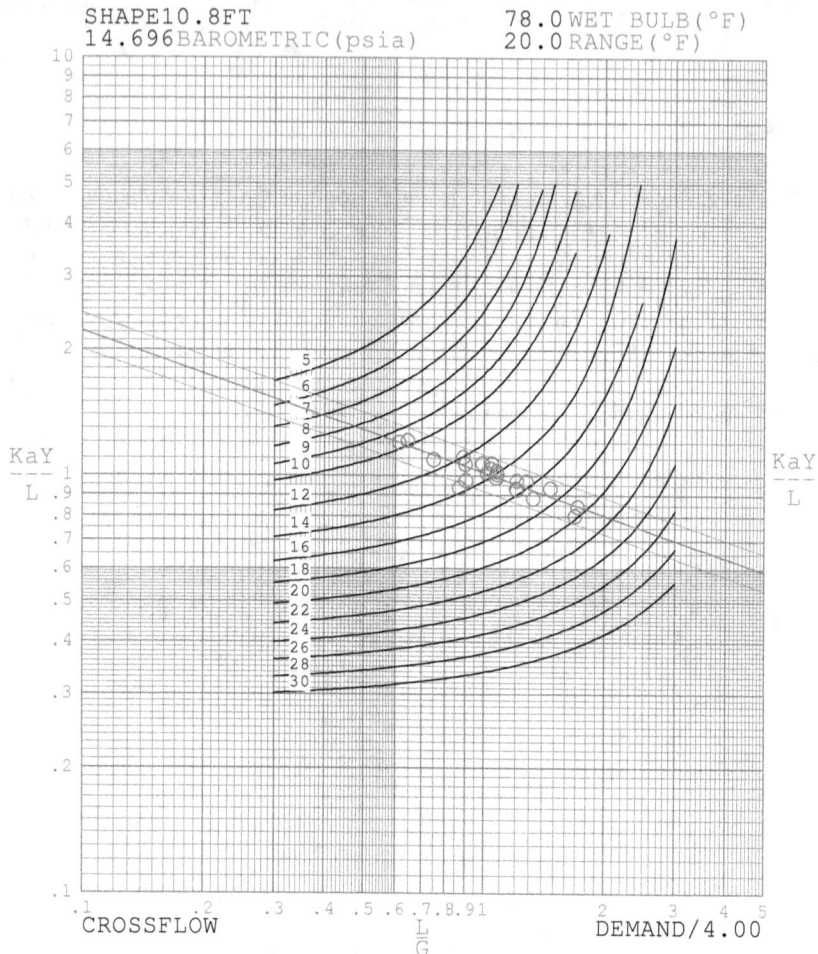

Figure 88. Typical Crossflow Curves

The program (DEMAND) works with both English and SI units. The input dialog is shown in this next figure:

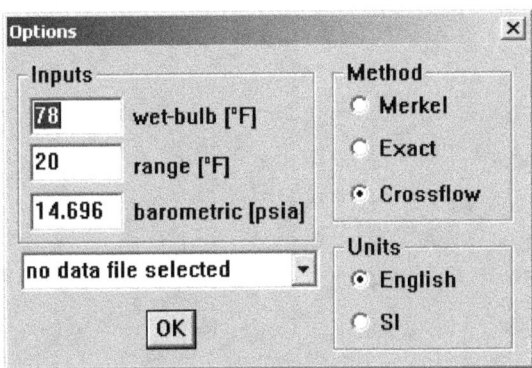

The resulting graph is pasted onto the clipboard. You can also paste the numerical values via the clipboard. If you have supply data, for instance from experiment or manufacturer, this can be read in from a file and a regression performed, which will be displayed on the graph with error bounds, as shown in red lines the preceding figure. The user-supplied data points are shown as red circles.

also by D. James Benton

3D Articulation: Using OpenGL, ISBN-9798596362480, Amazon, 2021 (book 3 in the 3D series).

3D Models in Motion Using OpenGL, ISBN-9798652987701, Amazon, 2020 (book 2 in the 3D series.

3D Rendering in Windows: How to display three-dimensional objects in Windows with and without OpenGL, ISBN-9781520339610, Amazon, 2016 (book 1 in the 3D series).

A Synergy of Short Stories: The whole may be greater than the sum of the parts, ISBN-9781520340319, Amazon, 2016.

Azeotropes: Behavior and Application, ISBN-9798609748997, Amazon, 2020.

bat-Elohim: Book 3 in the Little Star Trilogy, ISBN-9781686148682, Amazon, 2019.

Boilers: Performance and Testing, ISBN: 9798789062517, Amazon 2021.

Combined 3D Rendering Series: 3D Rendering in Windows®, 3D Models in Motion, and 3D Articulation, ISBN-9798484417032, Amazon, 2021.

Complex Variables: Practical Applications, ISBN-9781794250437, Amazon, 2019.

Compression & Encryption: Algorithms & Software, ISBN-9781081008826, Amazon, 2019.

Computational Fluid Dynamics: an Overview of Methods, ISBN-9781672393775, Amazon, 2019.

Computer Simulation of Power Systems: Programming Strategies and Practical Examples, ISBN-9781696218184, Amazon, 2019.

Contaminant Transport: A Numerical Approach, ISBN-9798461733216, Amazon, 2021.

CPUnleashed! Tapping Processor Speed, ISBN-9798421420361, Amazon, 2022.

Curve-Fitting: The Science and Art of Approximation, ISBN-9781520339542, Amazon, 2016.

Death by Tie: It was the best of ties. It was the worst of ties. It's what got him killed., ISBN-9798398745931, Amazon, 2023.

Differential Equations: Numerical Methods for Solving, ISBN-9781983004162, Amazon, 2018.

Equations of State: A Graphical Comparison, ISBN-9798843139520, Amazon, 2022.

Evaporative Cooling: The Science of Beating the Heat, ISBN-9781520913346, Amazon, 2017.

Forecasting: Extrapolation and Projection, ISBN-9798394019494, Amazon 2023.

Heat Engines: Thermodynamics, Cycles, & Performance Curves, ISBN-9798486886836, Amazon, 2021.

Heat Exchangers: Performance Prediction & Evaluation, ISBN-9781973589327, Amazon, 2017.

Heat Recovery Steam Generators: Thermal Design and Testing, ISBN-9781691029365, Amazon, 2019.

Heat Transfer: Heat Exchangers, Heat Recovery Steam Generators, & Cooling Towers, ISBN-9798487417831, Amazon, 2021.

Heat Transfer Examples: Practical Problems Solved, ISBN-9798390610763, Amazon, 2023.

The Kick-Start Murders: Visualize revenge, ISBN-9798759083375, Amazon, 2021.

Jamie2: Innocence is easily lost and cannot be restored, ISBN-9781520339375, Amazon, 2016-18.

Kyle Cooper Mysteries: Kick Start, Monte Carlo, and Waterfront Murders, ISBN-9798829365943, Amazon, 2022.

The Last Seraph: Sequel to Little Star, ISBN-9781726802253, Amazon, 2018.

Little Star: God doesn't do things the way we expect Him to. He's better than that! ISBN-9781520338903, Amazon, 2015-17.

Living Math: Seeing mathematics in every day life (and appreciating it more too), ISBN-9781520336992, Amazon, 2016.

Lost Cause: If only history could be changed..., ISBN-9781521173770, Amazon, 2017.

Mill Town Destiny: The Hand of Providence brought them together to rescue the mill, the town, and each other, ISBN-9781520864679, Amazon, 2017.

Monte Carlo Murders: Who Killed Who and Why, ISBN-9798829341848, Amazon, 2022.

Monte Carlo Simulation: The Art of Random Process Characterization, ISBN-9781980577874, Amazon, 2018.

Nonlinear Equations: Numerical Methods for Solving, ISBN-9781717767318, Amazon, 2018.

Numerical Calculus: Differentiation and Integration, ISBN-9781980680901, Amazon, 2018.

Numerical Methods: Nonlinear Equations, Numerical Calculus, & Differential Equations, ISBN-9798486246845, Amazon, 2021.

Orthogonal Functions: The Many Uses of, ISBN-9781719876162, Amazon, 2018.

Overwhelming Evidence: A Pilgrimage, ISBN-9798515642211, Amazon, 2021.

Particle Tracking: Computational Strategies and Diverse Examples, ISBN-9781692512651, Amazon, 2019.

Plumes: Delineation & Transport, ISBN-9781702292771, Amazon, 2019.

Power Plant Performance Curves: for Testing and Dispatch, ISBN-9798640192698, Amazon, 2020.

Practical Linear Algebra: Principles & Software, ISBN-9798860910584, Amazon, 2023.

Props, Fans, & Pumps: Design & Performance, ISBN-9798645391195, Amazon, 2020.

Remediation: Contaminant Transport, Particle Tracking, & Plumes, ISBN-9798485651190, Amazon, 2021.

ROFL: Rolling on the Floor Laughing, ISBN-9781973300007, Amazon, 2017.

Seminole Rain: You don't choose destiny. It chooses you, ISBN-9798668502196, Amazon, 2020.

Septillionth: 1 in 10^{24}, ISBN-9798410762472, Amazon, 2022.

Software Development: Targeted Applications, ISBN-9798850653989, Amazon, 2023.

Software Recipes: Proven Tools, ISBN-9798815229556, Amazon, 2022.

Steam 2020: to 150 GPa and 6000 K, ISBN-9798634643830, Amazon, 2020.

Thermochemical Reactions: Numerical Solutions, ISBN-9781073417872, Amazon, 2019.

Thermodynamic and Transport Properties of Fluids, ISBN-9781092120845, Amazon, 2019.

Thermodynamic Cycles: Effective Modeling Strategies for Software Development, ISBN-9781070934372, Amazon, 2019.

Thermodynamics - Theory & Practice: The science of energy and power, ISBN-9781520339795, Amazon, 2016.

Version-Independent Programming: Code Development Guidelines for the Windows® Operating System, ISBN-9781520339146, Amazon, 2016.

The Waterfront Murders: As you sow, so shall you reap, ISBN-9798611314500, Amazon, 2020.

Weather Data: Where To Get It and How To Process It, ISBN-9798868037894, Amazon, 2023.